Structured Questions for GCSE Physics

92

TONY YORK
BSc, ARCS
Head of Physics,
Alleyn's School, London

HODDER AND STOUGHTON
LONDON SYDNEY AUCKLAND TORONTO

CONTENTS

Preface

PREFACE

In their sample papers for GCSE, each of the examination boards has included questions which provide information about a situation, and then ask questions on it. The questions in this book are of this same 'comprehension' type. The information is presented as a passage, diagram, table or graph, or some suitable combination of these. The questions which follow are structured to test knowledge or understanding of particular aspects of the situation. Where the situation is complex, the structure leads the student through an argument in easy stages.

The questions are more extensive than their examination equivalents, so that one question may be used as a complete exercise, for tests, revision or homework.

The vast majority of the questions are based on everyday, medical or technological applications of physics. This is specifically intended to meet the aims set out in the National Criteria for physics:

2.2 To promote awareness and understanding of the social, economic, environmental and other applications of physics.

2.4 To provide a basic knowledge and understanding of . . . the applications of physics which contribute to the quality of life in a technologically based society.

It is hoped that they will give teachers and students many 'real life' examples to make their teaching and learning more interesting and relevant.

Since the questions are about real situations, they often contain ideas from different topics within physics. The divisions in the Contents are therefore not in any way rigid, and are only included for convenience, as teaching may still follow these divisions.

I would like to thank Chris Marvin for providing the basis of 10 electronics questions, John Clarke for proof-reading the typescript, and Kate York for checking and advising on the language level.

Tony York

Front cover shows optical fibres (Telefocus: a British Telecom photograph)

1   Lighting in a doll's house

A hobby magazine publishes two circuit diagrams for lighting a model house. Each one has eight bulbs, four downstairs and four upstairs. One circuit uses torch bulbs which have '3 V, 0.2 A' written on them and cost 20 p each. The other one uses car side-lamp bulbs which have '12 V, 4 W' written on them and cost 40 p each. The cost of the holders is the same for each type of bulb. Both circuits use the same sort of wire and switches. The switches cost 40 p each. Both circuits are operated by a 12 V transformer which runs off the 240 V mains supply.

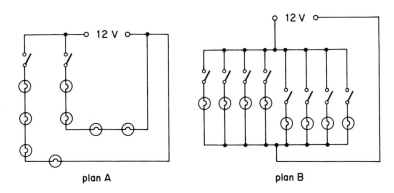

plan A   plan B

(a) If the bulbs are to light with their normal brightness, which type
     is used in each circuit?   *(1)*
(b) What is the power of a torch bulb?   *(2)*
(c) How much current flows through each car bulb?   *(2)*
(d) Calculate the resistance of each type of bulb.   *(2)*
(e) Plan B is more expensive to install than Plan A.
       (i) Calculate the *difference* in cost of the bulbs and switches
            for A and B.
      (ii) Give one other reason why Plan B is more expensive to
            install.
     (iii) Give two advantages of Plan B.   *(5)*
(f)   (i) Calculate the total current provided by the 12 V supply in
            each circuit.
      (ii) How many times more expensive is Plan B to run than
            Plan A?   *(3)*

Total 15 marks

2 Fuse investigation

A physics student measures how much current is needed to 'blow' (i.e. melt) various fuses. Some of the results are shown below.

(i) Different lengths of the same wire:

length/mm	5	10	15	20	25	30
current causing 'blow'/A	2.9	2.6	2.4	2.3	2.3	2.3

(ii) Different diameters of wire, same material and same length:

diameter/mm	0.25	0.50	0.75	1.00
current causing 'blow'/A	2.0	5.7	10.4	16

(iii) Effect of placing same length of same wire in glass tube:

without tube, current needed 2.3 A
with tube,   current needed 1.8 A

(a) Suggest suitable instruments to measure the
    (i) current through the wires
    (ii) length of wire
    (iii) diameters of the wires.   *(3)*
(b) The student explains table (i) by conduction of heat away from the ends of the wire to the crocodile clips holding the ends.
    (i) How does this theory explain why a larger current is needed to melt short lengths of wire?
    (ii) How does it explain why the current needed to melt the wire is the same for different, long, lengths of wire?   *(3)*
(c) Suggest your own theory for the results in table (iii).   *(2)*
The results in table (ii) are difficult to explain. See if you can answer the following questions provided by his teacher to help him.
(d) If the diameter of the wire doubles, what happens to the
    (i) surface area
    (ii) energy needed to keep the wire at the same temperature
    (iii) product I^2R
    (iv) resistance R
    (v) current I?   *(5)*
(e) Check whether your answer to (d) (v) agrees with the table. Use the results for diameters 0.25 and 0.5 mm, and 0.5 and 1.0 mm.   *(2)*

Total 15 marks

3 Energy uses in the home

Everyone who runs a home has many bills to pay, and these include large bills for energy. Many homes have both electricity and gas supplies. Most people use their electricity supply for some things, even if they also use gas, oil or coal for heating. The normal cost of electrical energy is more than the cost of some other types, e.g. gas. It is possible to use 'cheap rate' or 'off peak' electricity for water heating and room heating. It is not easy for a householder to compare the cost of different types of energy, because different units are used. Electricity boards use the 'kilowatt-hour' (3.6 MJ) which cost 5.3 p in 1986, but gas boards use the 'therm' (105 MJ approx.) which cost 38 p in 1986.

(a) Calculate the cost of each MJ (megajoule) of energy for
    (i) electricity
    (ii) gas. *(2)*

(b) (i) List 6 different household devices which use electricity.
    (ii) Write down what form of energy the electrical energy has been converted to in each one.
    (iii) Say for each one whether or not it would be easy to use some other form of energy to do the job.
    (iv) Why is it worth using electricity for some things, even when it is so expensive? *(10)*

(c) (i) What is meant by 'off peak' electricity?
    (ii) How can water heated using off peak electricity be kept hot until it is needed?
    (iii) How can energy produced at off peak times be stored for room heating later? *(4)*

(d) Name 2 ways of stopping heat energy escaping from a house. *(2)*

(e) Calculate the cost of
    (i) a 100 W bulb switched on for 10 h
    (ii) a 1.5 kW electric fire switched on for 4 h
    (iii) a 2.5 kW electric kettle switched on for 3 min. *(3)*

(f) A particular house has various electrical devices which are never switched off (e.g. clock, central heating time switch etc.). Altogether, these draw a current of 0.1 A from the 240 V supply. Calculate the cost of this per week. *(2)*

(g) Find out the cost of a kilowatt-hour ('unit') and of a therm at present by looking at your electricity and gas bills. *(2)*

Total 25 marks

4 Electricity transmission

Power stations generate alternating current (a.c.) so that the voltage can easily be changed using transformers. Very high voltages are used because less energy is wasted in heating up the cables taking the current from power station to consumer. The cables may be supported from pylons or towers, or they may run underground. Putting cables underground may cost twenty times as much as using pylons. The table shows the properties of metals used in electrical cables:

metal	mass/g of 1 metre	cost	strength	electrical conductor
copper	180	high	low	very good
aluminium	50	medium	low	good
steel	150	low	high	medium

It is difficult and expensive to keep stopping and starting power stations. Most power stations are therefore kept running all the time. Some of the 'off peak' electricity is used in 'pumped storage stations'. Water is pumped up into a high reservoir during the night, and then used to generate electricity at peak times.

(a) Explain briefly how a transformer works. *(3)*

(b) (i) If the voltage used is doubled, what happens to the current, for a fixed amount of power supplied?

    (ii) Because of the current change in (i), by what factor is the power wasted in the cables reduced? *(2)*

(c) (i) Why is it much more expensive to use underground cables?

    (ii) State two places where cables would be put underground in spite of the cost. *(3)*

(d) In a pumped storage station,

    (i) what form of energy is given to the water to raise it to the high reservoir?

    (ii) list the energy changes that take place when the water flows back down from the reservoir. *(4)*

(e) (i) Why is aluminium the best choice for overhead cables?

    (ii) Why do aluminium cables used overhead have steel wires running through them?

    (iii) Why is copper used for underground cables? *(3)*

Total 15 marks

5 Car lighting system

The diagram shows part of a car electrical system, which operates some of the lights.

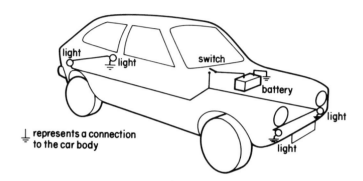

(a) The diagram only shows one lead from the battery to each bulb, but a complete ciruit must have two. How does the current get back to the battery? *(2)*

(b) If a car owner adds extra lights to the car, why must the paint be removed from the body-work where the electrical contact is made? *(1)*

(c) The two rear lights are wired up incorrectly, and only light faintly.
  (i) Explain why they only light faintly.
  (ii) Draw a diagram to show how the wires should be connected to make the lights work normally. *(2)*

(d) There are four side lights, and a number plate light. Each bulb is labelled '12 V, 4 W'.
  (i) What does the label mean?
  (ii) How much current flows through each bulb (assuming it is connected properly!)?
  (iii) How much current does the battery supply to these bulbs altogether? *(3)*

(e) If the lights are on, they go dim when the car is started. Why is this? *(2)*

Total 10 marks

6 Electricity supply and demand

The demand for electrical energy varies enormously during each
24 hours. The sort of typical variation for a winter's day is shown in the
graph below.
There are three main types of power station used to supply the energy.
The cheapest and most efficient are nuclear power stations. These may
be 40% efficient. They are built well away from heavily populated areas.
Next in cost are the coal fired power stations. The most costly are the oil
fired power stations. A small oil fired station near to a populated area
may only be 30% efficient. However some of the waste energy, in the
form of hot water, can be used to heat nearby hospitals and factories.
Because electrical energy is so difficult to store, the generating boards
have built 'pumped storage' stations. In these electrical energy is used to
pump water uphill to a high reservoir. When extra energy is needed the
water runs back downhill and drives generators. This process is about
75% efficient.

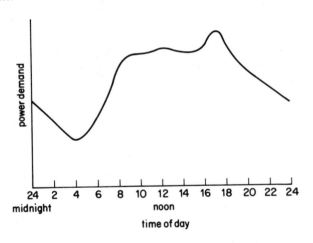

(a) Why is there a peak in the demand between 16.00 and
    18.00 hrs: (Remember it is a winter's day.)   (2)
(b) What percentage of the energy from the original fuel is wasted
    in
    (i) a nuclear power station
    (ii) an oil fired power station (if no hot water is used for
         heating)?   (2)

(c) In what form is this energy wasted? *(1)*

(d) The figures quoted are for converting the energy from the fuel into *electrical* energy. Explain why the oil fired station may have an *overall* efficiency greater than 30%. *(3)*

(e) Explain why nuclear power stations are built a long way from large populations. *(2)*

(f) Why can't the overall efficiency of a nuclear power station be increased by using hot water to heat hospitals and factories? *(2)*

(g) At what time of day do you think the electricity boards pump water uphill in a pumped storage station? *(2)*

(h) At what time of day do you think the electricity boards run the water back downhill? *(2)*

(i) For each megajoule (million joules) of energy originally in the form of nuclear fuel, calculate
    (i) the electrical energy generated
    (ii) the electrical energy finally available if the pumped storage system is also used. *(4)*

Total 20 marks

7 Tanker corrosion warning

Road tankers carry tons of liquid from place to place. The metal tank is lined with a material like polythene, which doesn't react chemically with most liquids. Because the road tanker bumps and vibrates as it drives along, the lining may crack. A dangerous liquid, like an acid, may then corrode the metal of the tank, and there might be a leak. The diagram shows a possible warning device. If liquid leaks through a crack in the polythene it touches the metal tank. An electric current can then flow from the circuit at A, through the liquid, through the metal tank, and back to the circuit at B. The current can be used to turn on an alarm, and the lining can be replaced before a dangerous leak occurs.

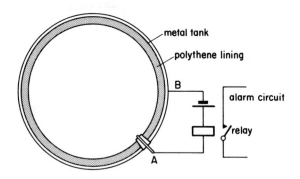

(a) What is the name for materials which
    (i) allow a current to flow through them
    (ii) do not allow a current to flow through them?   *(2)*
(b) For the alarm to work, explain why the metals must allow a current to flow through them, but polythene must not.   *(2)*
(c) The alarm will only work for some liquids. Say what the liquids must do if they will make the alarm work.   *(2)*
(d) Will the alarm work for a petrol tanker? Explain your answer. *(2)*
(e) The alarm will work for tankers carrying acid. What happens when an electric current passes through an acid?   *(2)*

Total 10 marks

8 Detecting oil and gas leaks

When drilling for oil and gas there is a danger of leaks. If these are large, they can cause very serious damage to the environment. Oil companies use scores of gas detectors on their oil platforms. (An oil leak also produces enough gas to be detected.) A gas detector is shown below:

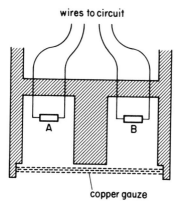

A and B are identical resistors. B is coated with a reactive chemical, and any flammable gas which comes into contact with it burns. This raises the temperature of B and makes its resistance change. The resistance of A does not change. The *difference* in resistance is used to set off an alarm.

The burning gas could cause a fire. To prevent this the resistors are behind a fine copper gauze. This stops the heat from the burning gas reaching the rest of the gas outside the detector, and so stops it from catching fire.

(a) Will the resistance of B increase or decrease as it gets warmer if it is made of a semi-conductor? *(1)*

(b) On a North Sea oil rig the temperature changes quite a lot just because of the weather. If B were used on its own, without A, its temperature might change enough to set off the alarm when there was no gas. Explain how using A and detecting the *difference* in resistance avoids this false alarm. *(3)*

(c) How does the *copper* gauze stop heat from the burning gas reaching the gas outside the detector? *(2)*

(d) Give two ways in which large oil spills damage the environment. *(4)*

Total 10 marks

9 Resistance transducers

A *transducer* is a device which changes one sort of energy to another. A well known example is a microphone which converts sound energy into electrical energy.

Resistance transducers produce a change in electrical resistance so that some other physical quantity can be measured electrically. Some examples are given below:-

1 Position

The movement or position of part of a machine can be used to move a sliding contact on a fixed resistor. This is illustrated below. The fixed resistor can also be bent in a curve, so that the movement measured can be a turning movement. (See question number 63 for an example of a position transducer in use.)

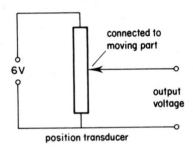

position transducer

2 Temperature

Temperature can be measured using a fine piece of metal wire or a thermistor. A thermistor is a piece of semi-conductor material. They can both be used to measure temperature, because their resistance changes as they get hot or cold.

3 Gas flow

A thermistor which has a current flowing through it will be warmed up by the current. If it is placed in a pipe which has gas flowing through it, the gas flow will cool the thermistor down. The change of resistance produced is a measure of the speed of gas flow in the pipe.

4 Light

A light dependent resistor (LDR) is made of semi-conductor material. When light falls on it, its resistance decreases. It can therefore be used to measure light levels.

5 Strain (Stretching)

A strain gauge consists of a thin piece of wire which might be firmly glued to a girder in a bridge. When a heavy load crosses the bridge, the girder may stretch slightly. The wire strain gauge also stretches, and so its resistance changes. This change in resistance can be used to measure the strain that the bridge is under.

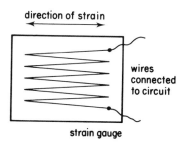

direction of strain

wires connected to circuit

strain gauge

(a) What is the output voltage from the position transducer when the slider is
   (i) at the top of the resistor
   (ii) ¼ of the way down from the top
   (iii) at the bottom of the resistor?   *(3)*

(b) Sketch a position transducer which has been altered so that it is operated by a turning movement.   *(2)*

(c)   (i) If a metal wire gets hot, does its resistance increase or decrease?
   (ii) If a thermistor gets hot, does its resistance increase or decrease? Explain why this change takes place.   *(3)*

(d)   (i) Explain why a gas flowing past a thermistor will cool it.
   (ii) Which will cool the thermistor most, a slow gas flow or a fast gas flow?
   (iii) Provide the missing word in the following sentence. 'As the gas flow gets faster, the resistance of the themistor . . .'   *(4)*

(e) Name one device or situation where an LDR is used to measure light or for control.   *(2)*

(f) In a strain gauge, how does stretching alter
   (i) the length of the wire
   (ii) the thickness of the wire
   (iii) How do each of (i) and (ii) alter the resistance of the wire?
   (iv) Suggest why the strain gauge is a zig-zag shape.   *(6)*

Total 20 marks

10 Using resistance to measure abrasion

As magnetic recording tape passes at high speed over the recording and play-back heads of a tape recorder it causes wear. A way of testing tape to see how much wear it causes is described below. The tape passes over a ceramic cylinder coated with a thin layer of resistance material. (A ceramic substance is like pottery.) As the tape wears away the resistance material its resistance changes. This resistance can be measured, and so the wear caused by the tape calculated.

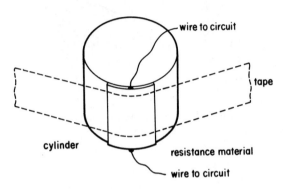

(a) Why is the cylinder not made of metal? *(1)*
(b) Draw and label a circuit diagram showing how you would measure the resistance. Include how to calculate the resistance from any measurements you would make. *(4)*
(c) How does the resistance change as the thin layer is worn away? *(1)*
(d) What effect does this change in resistance have on the current flow from a fixed voltage battery? *(1)*
(e) In practice, the resistance is measured electronically by passing a fixed *current* through the layer and measuring the voltage across it. How does the voltage change as the layer wears away? *(2)*
(f) Why must the current mentioned in (e) be small? *(2)*
(g) When a test of some tape is done, the amount of wear increases if the atmosphere is damp. Suggest a reason for this. *(2)*
(h) Why does the device have to be replaced occasionally? *(2)*

Total 15 marks

12

11 Electric motor

The diagram on the left shows the basic principle of an electric motor. It is the sort of diagram which is common in physics text books. The diagram on the right shows a real motor used in a vacuum cleaner.

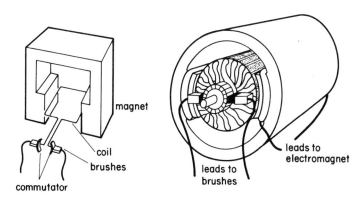

(a) The basic motor only has one coil, which has to 'free wheel' to get the connections round from one side of the commutator to the other. Why does the real motor have many coils? *(2)*

(b) Why does the real motor have many sections on the commutator? *(2)*

(c) Why are the coils of the real motor wound on a cylinder of iron? *(2)*

(d) Why is the cylinder of iron made from thin sheets of iron (called laminations) rather than a solid piece? *(2)*

(e) Would the basic motor work if alternating current was used? Explain your answer. *(2)*

(f) The real motor uses an electromagnet, not a permanent one. Explain how this allows the motor to be used with alternating current. *(3)*

(g) Why is the iron of the electromagnet also made of laminations? *(2)*

Total 15 marks

13

12 Magnetic paint thickness meter

A student read about a simple magnetic instrument for measuring the thickness of paint. She decided to base a physics investigation on it. She measured the force needed to pull the magnet away from the steel sheet on the spring balance. The spring balance was gradually stretched by gently turning the screw of the clamp. The next stage was to put a single sheet of very thin paper between the magnet and the steel sheet, and measure the force again. Then she used two sheets of paper, then three etc. A table of her results is shown.

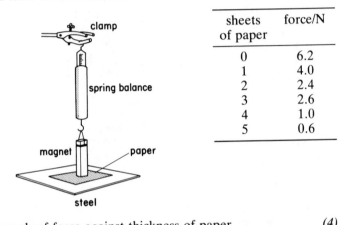

sheets of paper	force/N
0	6.2
1	4.0
2	2.4
3	2.6
4	1.0
5	0.6

(a) Plot a graph of force against thickness of paper. *(4)*
(b) One of the results looks suspect. Say which one, and suggest how a mistake might have been made. *(2)*
(c) Why does the force needed to pull the magnet away decrease as the thickness of paper increases? *(1)*
(d) How would you measure the thickness of the paper? *(2)*
(e) Sketch the shape of the magnetic field near one pole of the magnet when it is
    (i) well away from the steel sheet
    (ii) near to the steel sheet. *(4)*
(f) Layers of paint may be much thinner than sheets of paper. A friend suggests that aluminium leaf could be used as the separating material, to get smaller separations. Someone else says that a metal should not be used. Explain why you think that aluminium is, or is not, suitable. *(2)*

Total 15 marks

13 Magnetic impact meter

Accidents to various sorts of vehicles do happen from time to time. The designers try to build safety into their vehicles. With large vehicles, like hovercraft, models are used in testing. A simple application of physics is used to find out how much force is produced in a crash. A strong magnet has small tubes of various lengths fixed to it. Each tube has an adjustable brass screw in it. A steel ball bearing rests on top of the screw. It is held in place by the magnetic field. The field gets weaker further away from the magnet.

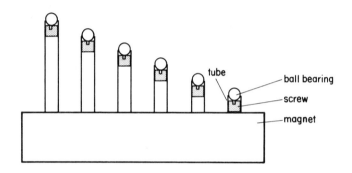

The magnet is put inside the model. When the model crashes, the force produced will knock some of the ball bearings off. It is possible to tell just how big the force was by seeing how many ball bearings have been knocked off.

(a) How does the force of attraction on the ball bearings towards the magnet change as they get further away? *(1)*

(b) Which ball bearings are easiest to knock off? *(1)*

(c) Why are the screws made of brass, but the ball bearings are made of steel? *(3)*

(d) Name a suitable material for the tubes. Give a reason why it is suitable. *(3)*

(e) The magnet does *not* have its poles at each end, but on the top and bottom faces. Explain why this is a good idea. *(2)*

Total 10 marks

14 Magnetic crack detection

If very small cracks in metals can be detected before they become large, many accidents can be avoided. For steel pipes a method has been developed using magnetism. A coil of wire is wrapped round the pipe, and a current passed through the coil. This magnetises the pipe and any cracks in the direction shown in the diagram can be found by sprinkling iron filings on the pipe. The filings cluster round the crack and so show up where it is. Cracks *along* the pipe do not show up. To find any cracks along the pipe, a magnetic field in a different direction must be produced.

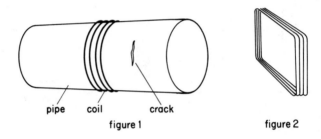

pipe   coil   crack

figure 1   figure 2

(a) Why can't the method be used for copper pipes? *(1)*
(b) Would the method be useful for detecting small cracks in aircraft? Explain your answer. *(3)*
(c) In which direction is the magnetic field in the pipe? *(2)*
(d) What happens if the current in the coil is increased? *(1)*
(e) What happens if the current in the coil is reversed? *(1)*
(f) Why do iron filings cluster round the crack? *(2)*
(g) Draw a copy of the coil shown in figure 2, and sketch the shape of the magnetic field it produces when a current flows through it, but it is not near any iron. *(2)*
(h) The crack only shows up if it is *across* the direction of the field. How would you place the coil in (f) so that the magnetic field it produces will show up cracks which are *along* the pipe? *(3)*

Total 15 marks

15 'Ding-dong' door bell

A 'ding-dong' type of door chime is represented in the following diagram:

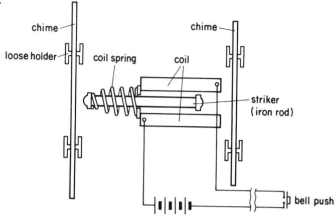

(a) What happens to the coil of wire (solenoid) when the door bell is pushed? *(1)*
(b) Why is the rod inside the coil made of iron? *(1)*
(c) What happens to the iron rod when you *stop* pushing the bell? *(1)*
(d) What is the coil spring for? *(1)*
(e) What would happen, and what would you hear, if you pushed the bell but didn't let go again? *(2)*
(f) Why would you be unpopular if you made a habit of ringing the bell as described in (e)? *(1)*
(g) Why are the 'chimes'
   (i) only loosely held on their supports
   (ii) of different lengths? *(2)*
(h) After some decorating has been done the chimes no longer work. Make sensible suggestions to decide if
   (i) the bell push has stopped making contact. (Maybe paint got in)
   (ii) the wire from the bell push to the chime has broken. (This may not be obvious because it may have broken inside the plastic insulation)
   (iii) the batteries are flat
   (iv) the thin wire of the coil has broken. *(6)*

Total 15 marks

16 Electromagnetic flowmeter

Many chemical industries use miles and miles of pipes with liquids flowing through them. For many of these it is important to know how much liquid flows. One measuring method uses electromagnetic induction. The flow of liquid through the pipe makes a turbine wheel (like a propeller) turn. The wheel has many magnets arranged round its edge. As it turns, these magnets pass by a coil of wire wound on a piece of iron. Pulses of voltage are generated across the coil. The faster the liquid flows through the pipe the faster the wheel turns, and more pulses are produced each second. If the electric pulses are counted, the speed of liquid through the pipe can be measured.

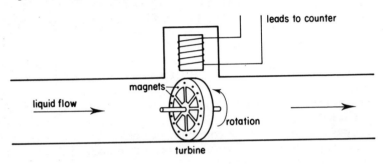

(a) Why is the coil of wire wound on a piece of iron? *(1)*
(b) In which direction should the magnetic field in the coil be to produce the biggest effect? *(1)*
(c) What is the effect of winding more turns of wire on the coil? *(1)*
(d) Why is no voltage produced if the wheel is stationary? *(1)*
(e) Why will placing each magnet the other way round from the one next to it produce a bigger effect? *(2)*
(f) The passage says that faster flow produces more pulses of voltage each second. What other effect does faster flow have on the voltage? Explain your answer. *(2)*
(g) How will placing more magnets round the rim of the wheel affect the pulses of voltage? *(2)*
(h) Should the wheel be made of magnetic material or not? Explain your answer. *(2)*
(i) This type of measurement is safe for dangerous liquids such as oil, because it is totally enclosed. However, there are disadvantages. Try to think of one, and explain why it is a disadvantage. *(3)*

Total 15 marks

17 Help for the hard of hearing

It is becoming much more common for handicapped children to go to ordinary schools, rather than special ones. The schools they go to may need special equipment to help them. Some schools have a 'hearing unit' so that they can teach children who are almost deaf in a class of children with normal hearing. The teacher speaks normally to the class, but also wears a small microphone. This converts the sound signal into tiny electric currents. These currents are amplified, and then fed to a loop of wire which goes round the walls of the room at waist height. Because these currents are changing they produce a changing magnetic field. The deaf children in the class have a small loop of wire attached to their hearing aids. These loops of wire will detect the changes in magnetic field by producing tiny currents. The hearing aid uses these to help the deaf person to hear. The hearing aid can be switched back to picking up normal sound.

The same principle can be used for listening to radio or television programmes. In an ordinary T.V. set the sound is produced by electric currents fed to a loudspeaker. In a modified set these currents are also amplified and fed to the loop of wire round the room. Low notes are amplified more than high ones. The volume of sound can be normal, so the people with normal hearing and those who are deaf can listen to the same programmes without interfering with each other.

(a) Sketch the shape of magnetic field produced by a current
    flowing through a loop of wire round a square room. *(2)*
(b) State one advantage and one disadvantage of using a lot of turns
    of wire for the loop round the room. *(3)*
(c) The coil on the hearing aid can be altered in various ways. Say
    what effect each of the following changes will have on the signal
    picked up.
    (i) More turns of wire are used.
    (ii) The coil has a larger area.
    (iii) The axis of the coil (i.e. a line through the middle of the
        coil), which is normally in the same direction as the
        magnetic field, is turned through 90° *(4)*
(d) What difference will it make to the signal picked up if
    (i) a steady current flows through the room loop
    (ii) the current in the room loop is changing slowly
    (iii) the current in the room loop is changing quickly? *(3)*
(e) Low notes are produced by slowly changing currents, high
    ones by quickly changing currents. Why do the low notes need
    to be amplified more than the high ones? *(3)*

Total 15 marks

18 Relay

An automatic washing machine switches on and off at various times during the wash cycle. Only small currents (far too small to operate the motor, pump etc.) are provided by the automatic control device. There are many other situations when a large current needs to be switched on by a much smaller one. A device designed to do this is the relay, as shown below. A small current flows through the coil from A to B. This magnetises the iron core, which attracts the armature, and so closes the spring contacts. This switches on the large current for the motor.

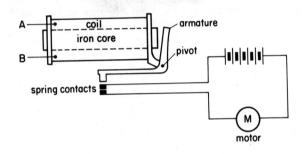

(a) How is the magnetic field in the iron core affected by
   (i) passing more current from A to B
   (ii) winding more turns on the coil? (2)
(b) What eventually happens to the magnetic field if you go on increasing the current and the number of turns? (2)
(c) What sort of material is used to make the armature? (2)
(d) The distance from the pivot to the contacts is longer than the other arm of the armature. What advantage does this give? (2)
(e) When the spring contacts are opened and closed, a spark may jump across the gap. What property(ies) should the material they are made of have to stop this causing any damage? (2)

Total 10 marks

19 Recording on magnetic tape

The tape used in tape recorders is coated with a layer of magnetic material. The sound to be recorded is converted into a pattern of changing electric current. This current is passed through the coil of wire. The metal ring is magnetised by the current. Some of the magnetic field passes through the coating on the tape and magnetises that in the same pattern. The pattern of the sound waves has now been converted into a matching magnetic pattern stored on the tape.

The same sort of 'head' can be used to 'play back' the tape. As the magnetised tape passes the head its magnetic field magnetises the metal. The changes in magnetic field induce currents in the coil. The currents can then be amplified and used to produce sound.

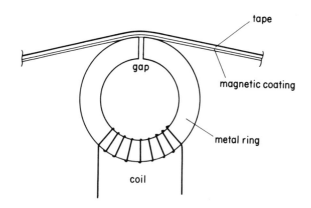

(a) (i) What device converts sound energy into electrical energy?
     (ii) What device converts electrical energy into sound energy? (2)
(b) The passage uses the words 'magnetic' and 'magnetised'. Explain the difference in meaning of these two words. (2)
(c) A magnetically 'soft' material can be magnetised easily, but also loses its magnetism easily. Explain why this sort of material is needed for the metal ring. (2)
(d) What happens to the magnetic field in the metal ring
     (i) if the current in the coil increases
     (ii) if the current in the coil reverses? (2)
(e) How would music sound if the tape went past the play-back head faster than it had gone past the recording head? (2)

Total 10 marks

20 Electrostatic crop spraying

Spraying fertilizer or insecticide on crops is not always very effective.
The liquid clings together (surface tension) making it difficult to
produce small drops. The spray may not spread very widely or evenly.
Drops may bounce off the leaves and be wasted on the ground. All of
these things mean that more chemical has to be used, increasing worries
about the environment and increasing the cost. There are now reliable,
cheap and portable high voltage generators available which make it
possible to overcome these problems. The drops are electrically charged
as they are sprayed. This makes it easier for small drops to form, the
spray spreads out more and this spread is more even. As the drops get
near the plants, the plants themselves become slightly charged and so
attract the drops. Making the drops charged means using extra energy.
The sort of voltage used is about 5 kV, and the amount of charge put on
the spray about 1 mC for each kilogram of liquid.

(a) All the drops are given the same sort of charge, i.e. all positive
   or all negative. Explain why this makes it easier for
   (i) the liquid to break up into small drops
   (ii) the spray to spread out more
   (iii) the spray to be more even, i.e. the drops are not sprayed
        in some concentrated regions with some gaps. (3)
(b) Give two reasons why it is an advantage to have small drops. (2)
(c) The plants are reasonable conductors of electricity, and are
   connected to 'earth'. If the drops are positively charged,
   (i) what sort of charge will there be on the plants
   (ii) where does this charge come from, and why? (3)
(d) Calculate how much energy in joules is needed to charge each
   kilogram of liquid. (2)
(e) The cost of 3 600 000 J of electrical energy bought from an
   electricity board is about 5 p. How many kilograms of spray
   could be charged for this cost? (2)
(f) Why is it unlikely that a farmer would actually buy the energy
   needed from an electricity board? (2)
(g) In what form would the farmer buy the energy needed? (1)

Total 15 marks

21 Ink jet printing

Fast, high quality printing can be achieved by spraying ink directly on to the paper. The ink is passed through a 'piezo-electric' tube. When an electrical pulse is applied to the tube it changes size. This is done 100 000 times each second, and as a result the ink comes out in 100 000 tiny drops each second. As the drops come out, they pass through a metal ring which has a voltage applied to it. This puts a charge on each drop. If the voltage is changed quickly enough, each drop can have a different charge put on it. The charged drops then pass between a pair of metal plates with a fixed voltage across them. Any drops with no charge hit the shield and do not reach the paper. Drops which are charged are deflected, and hit the paper at different places, depending on their charge, and so build up any required shape on the paper.

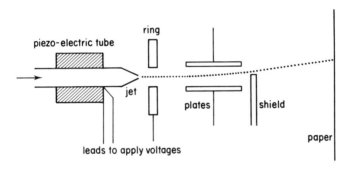

(a) What path will an uncharged drop take as it passes between the deflecting plates? (2)

(b) If the drops are given a negative charge, how should the voltage be applied to the deflecting plates? (2)

(c) How are the drops directed to the top of a line of printing, say to the dot above an 'i'? (2)

(d) How are the drops directed to the bottom of a line of printing, say to a full stop? (2)

(e) Why can a 'non contact' method like this be made very fast? (2)

Total 10 marks

23

22 Dielectric heating

'Dielectric' is a name for some types of electrical insulators, for example polythene and nylon. In the molecules of these insulators, the electric charge of the electrons and nuclei is not distributed evenly. This means that one end of the molecule is positive and the other end negative. This type of molecule is called a polar molecule. If a piece of dielectric is put between two metal plates connected to a high voltage supply, the molecules try to line up. If the high voltage is *alternating,* the molecules try to line up first one way, then the other. This means that the molecules are being made to vibrate rapidly, and the material heats up. Two pieces of material squeezed together between the metal plates can be heated so that they soften and join together. This sort of joining, or welding, is widely used for packaging materials and for the seams in waterproof clothing.

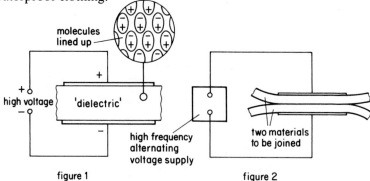

figure 1   figure 2

(a) Which end of a polar molecule has the greatest concentration of electrons? *(1)*
(b) What sort of electric charge does the nucleus of an atom have? *(1)*
(c) Why do the molecules try to line up in the direction shown in figure 1? *(1)*
(d) The word 'try' is used several times in the passage. Why are the molecules not free to line up entirely? *(2)*
(e) The voltage supply is labelled 'high frequency, alternating'.
    (i) What does alternating mean?
    (ii) What does high frequency mean?
    (iii) Why is a high frequency an advantage? *(3)*
(f) What does the power supply actually do to each metal plate to make it (i) negative, and (ii) positive? *(2)*

Total 10 marks

23 Ignition and control of gas flames

Gas is widely used for heating, in homes and in industry. Gas fires are often lit by an electric spark. The spark can be produced by squeezing a 'piezo-electric' material. When this is done a large voltage is generated, and this is what gives the spark. In some industrial applications, electricity is also used to detect if a gas flame goes out. As the gas burns, it produces many ions. Metal plates are put on each side of the flame. A high voltage is connected across the plates. A sensitive meter is included in the circuit. When the flame is burning, there is a flow of current. If the flame goes out, the current stops. When this happens the gas is automatically turned off.

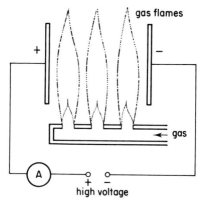

(a) A potential difference of 30 000 V is needed to make a spark jump across each centimetre of air. The spark gap in a gas fire ignition system is 4 mm. How much voltage must the piezo-electric material generate? *(2)*
(b) Where does the energy come from to produce the spark? *(1)*
(c) What is an *ion*? *(1)*
(d) The ions carry the current across the gap between the plates. Why do the ions move across the gap, and which way do they move? *(2)*
(e) What is the effect of increasing the voltage across the plates? *(1)*
(f) If the separation between the plates is 10 cm, why must the voltage be *less* than 300 000 V? *(1)*
(g) What carries the current through the wires? *(1)*
(h) What happens to positive ions if they touch the negative plate? *(1)*

Total 10 marks

24 Power supply for a home computer

A pupil has just bought a new home computer. It has a metal case. When he looks at the instruction manual he recognizes the circuit diagram for the power supply. This supply converts 240 V a.c. into a smaller direct voltage. The circuit diagram is shown below.

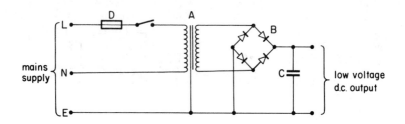

(a) What do the letters L, N. and E stand for in the mains supply? *(3)*
(b) What colour is each of the wires in (a)? *(3)*
(c) Not all electrical appliances use the wire marked E. Why is it essential in this computer?
(d) D is a fuse. Explain why it is in this circuit. *(2)*
(e) What is A, and what does it do? *(2)*
(f) B shows four diodes connected together to make a *bridge rectifier*. What does a *single* diode do? *(2)*
(g) What does the capacitor C do in this circuit? *(1)*

Total 15 marks

25 Amplifier

Many electronic devices use small voltages which need to be amplified. One possible way to do this is to use a transformer, shown below on the left. However, a transformer has some disadvantages, and so an amplifier may be used instead. This is shown on the right.

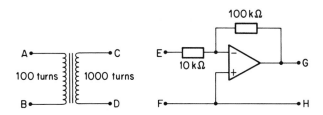

The following questions are about the ways in which the circuits are alike or different.

(a) A 1 volt battery is connected across A B and a voltmeter is connected across C D. What will the voltmeter read? Explain your answer. *(2)*

(b) The same battery is connected across E F and the voltmeter is connected across G H. What will it read? Explain your answer. *(2)*

(c) An alternating voltage with a peak volume of 1 V is connected across A B and then across E F.
  (i) What device would you use to examine the output voltage of each circuit?
  (ii) Describe the output voltages produced across C D and G H. *(5)*

(d) How can the output of the transformer be increased? *(1)*

(e) How can the output of the amplifier be increased? *(1)*

(f) The amplifier needs an external supply of energy. Give one way in which this can be an advantage, and one way in which it can be a disadvantage. *(4)*

Total 15 marks

26 Burglar alarm

Most burglar alarms use a bistable circuit for switching on and off. This means that if a burglar operates a hidden switch and the alarm is turned on, it *stays* on even if that switch does not stay turned on. The alarm can only be turned off using another switch, probably inside a locked cupboard.

The diagram shows a bistable circuit using two NAND gates (NAND gates are commonly available in cheap 'microchip' form.)

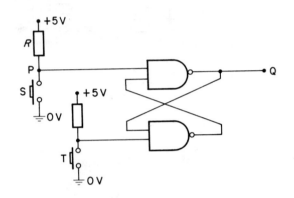

(a) Write down the truth table for
   (i) an AND gate
   (ii) a NAND gate. *(4)*
(b) Point P is at 5 V because of the connection through the resistor R. What happens to the voltage of point P when the push switch S is closed? *(1)*
(c) When switch S is pushed, the output voltage at Q becomes 5 V. What happens when switch S is released again? *(2)*
(d) Explain why this circuit is called a bistable. *(2)*
(e) Using the brief description at the beginning and your answers to the earlier questions, explain in more detail how the bistable can be used for a burglar alarm. You should say which switch is which, and what the 5 V output at Q is used to do. *(4)*
(f) A bistable is also a form of memory device used in computers. Explain how it is used. *(2)*

Total 15 marks

27 Light-operated relay

An electronics enthusiast has designed a circuit which switches car sidelights on when it starts to get dark. The circuit is shown below. It contains an NPN transistor, a light dependent resistor (LDR) and a resistor R. The transistor is used to switch on a relay, and the relay then turns the lights on.

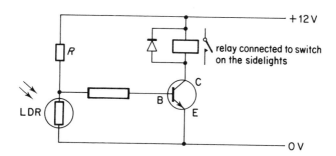

(a) The three terminals of the transistor are labelled C, B and E. What do these letters stand for? *(3)*

(b) When the LDR has light shining on it, its resistance is very much less than that of the resistor R. Explain why this turns the transistor off. *(3)*

(c) When the LDR is in darkness the transistor is switched on, that is current flows through it. Does this (conventional) current flow from E to C or from C to E? *(1)*

(d) Current through the transistor switches the relay on, so the car lights come on. A friend looking at the diagram says 'Wouldn't it be much simpler to have just the relay connected in series with the LDR, and not bother with the transistor? Explain why
  (i) the simple circuit would probably not work, and
  (ii) even if it did, it would switch the lights on in daylight and off when it got dark! *(5)*

(e) Why is there a diode connected across the coil of the relay? *(3)*

Total 15 marks

28 Safety switch

An engineer working in a factory has designed a circuit to act as a safety switch for a machine. If the machine operator's hand is in a dangerous position at the same time as the machine is turned on, an alarm will sound and the machine will be turned off. A block diagram of the system is shown below.

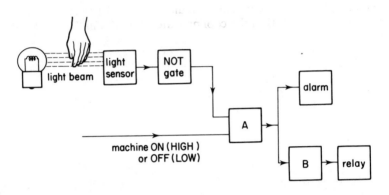

(a) What electrical device could be used as the light sensor? *(1)*
(b) The light sensor gives a LOW output when the light beam is broken. Explain what the NOT gate does. *(1)*
(c) The alarm is turned on by the device in box A. By looking at the conditions needed for the alarm to come on, decide what should be in box A. *(2)*
(d) The relay switches the machine on and off. Why is a relay a suitable device to do this? *(2)*
(e) The machine is switched ON when the relay is ON. It should be switched OFF when the output from A is HIGH. What must box B contain? *(2)*
(f) This switch, like most safety systems, is 'fail safe'. Explain what this means. *(2)*

Total 10 marks

29 Electronic thermometer

When a car engine is running normally, the cooling water keeps the temperature fairly steady. If something goes wrong, the engine may get hotter, and this could cause expensive and dangerous damage. A temperature gauge, or thermometer, gives the driver warning if this starts to happen. The fault can then be put right before too much damage is done.

The diagram shows a possible circuit. The thermistor is a semi-conductor device. It is in the cooling water, and the milliammeter is on the dashboard.

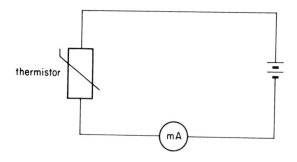

(a) Is the circuit a series one or a parallel one? *(1)*

(b)   (i) As the thermistor gets hotter, what happens to its resistance?

   (ii) What effect does this have on the milliammeter reading? *(2)*

(c) The current scale on the meter is removed and a scale in degrees celsius put in its place. Unfortunately the scale is not *linear*. What does this mean? *(2)*

(d) The thermistor becomes faulty and has to be replaced. The only replacement available has a lower resistance. How does this affect the current? *(1)*

(e) Give three possible ways of dealing with the problem in (d), and say which one you think would be the simplest and cheapest. *(4)*

Total 10 marks

30 Cathode ray oscilloscope (CRO)

The cathode ray oscilloscope is basically a voltmeter. However, it has several advantages over simpler sorts of voltmeters. One of the most important is that it can respond very quickly to changing voltages. It can also show a graph of voltage against time.

A CRO is used in many applications. A garage mechanic may use one for 'electronic engine tuning', there are many medical uses, and many sorts of engineers and scientists use them.

The questions below could be part of a training course for any of these people learning how a CRO works.

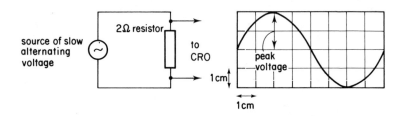

The diagram on the left shows a simple circuit. The diagram on the right shows what the trace on the oscilloscope looks like. The CRO controls are set so that the spot moves 1 cm upwards for every 1 volt applied. The 'time base' is set so that the spot covers 1 cm across the screen in a time of 0.1 s.

(a) What is the peak voltage shown on the screen? *(1)*
(b) What does the time base do? *(2)*
(c) How long does the alternating voltage take to go through one cycle? *(1)*
(d) What is the frequency of the alternating voltage? *(1)*
(e) What is the peak current flowing through the resistor? *(1)*
(f) A diode or rectifier is connected in series with the resistor. Sketch the new shape of the trace on the CRO screen. *(2)*
(g) What is the moving part in a CRO, and why can it respond very quickly? *(2)*

Total 10 marks

31 Oscilloscope in use

An oscilloscope is a voltmeter. It can show how a voltage changes with time. If connected to a suitable *transducer* it can show heart beats, breathing rates, or sound waves. (A transducer converts the energy of each of the things mentioned into a voltage. See question number 72 for an example.)

Figure 1 shows what the trace on the screen looks like from a heart beat, figure 2 from a baby breathing, and figure 3 from a musical note played on a recorder.

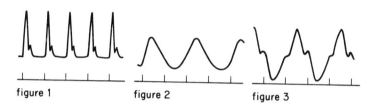

figure 1   figure 2   figure 3

In figures 1 and 2 the trace covers the screen at a rate of 1 division/s, but in figure 3 it is much faster, 1 division in only 1/1000 s.

(a) How many heart beats are there in one minute? *(1)*

(b) How would the trace alter if the person did some violent exercise and the heartbeat went up to 120 beats per minute? *(2)*

(c) How many breaths is the baby taking per minute? *(1)*

(d) How would a doctor or nurse be able to tell from the trace if the baby's breathing
  (i) slowed down
  (ii) speeded up? *(2)*

(e) Calculate the frequency of the note, i.e. how many oscillations there are per second. *(2)*

(f) Say how the trace from the recorder would change if it was played
  (i) louder
  (ii) at a higher pitch. *(2)*

Total 10 marks

33

32 Truth table

Truth tables originally had nothing to do with electronics, or even science. They were developed as a way of showing lots of complicated information in the study of logic. The sort of situation might be 'If a particular thing is true, AND something else is true, then a third thing must be true.' The idea has been adapted to the sort of 'logic' used in computers and electronics.
A logic circuit is shown below.

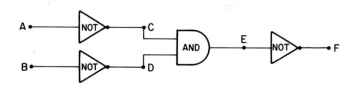

(a) What is a truth table (in the electronic sense)? (2)
(b) Write down the truth table for an AND gate. (2)
(c) The table below shows the different states at various places in the logic circuit in the diagram. Copy the table and fill in the blanks with a 1 or 0.

A	B	C	D	E	F
0	0	1	1	1	0
0	1	1		0	1
1	0		1	0	
1	1	0	0	0	

(2)
(d) Look at the output at F and compare it with the two inputs at A and B. What single logic gate could replace the whole circuit to give the same result at F? (2)
(e) The AND gate in the circuit is changed for an OR gate. Explain how this would alter the way the circuit behaves. (2)

Total 10 marks

34

33 Random number display

Many electronic games have some randomness built in. This means that they are not the same every time they are played, which stops them becoming very boring. The diagram below shows part of the design of such a game. When the button S is pushed and then released a number is displayed. It is chosen at random from 0 to 9.

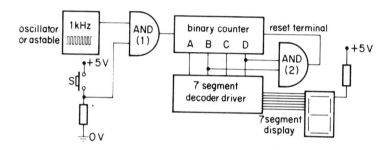

(a) The diagram shows a 1 kHz oscillator or astable. What does this do? *(2)*

(b) The first AND gate only allows the binary counter to count when the button S is pushed. Explain why an AND gate is used. *(3)*

(c) The counter produces a binary number at its terminals ABCD. What is a binary number? *(2)*

(d) The binary counter is reset at zero when the resent terminal becomes HIGH (logic state 1). Explain how the second AND gate resets the counter when it has counted to 10 (decimal!) *(3)*

(e) What does the decoder do? *(2)*

(f) The number (decimal) appears on the display each time the button S is released. Explain why it is a *random* number. *(3)*

Total 15 marks

34 OR logic gate with relay

Electronic logic often appears rather mysterious because it is usually done inside 'chips'. There are no moving parts, and nothing except small indicator lights or readings on meters to be seen. It is not always realised that the same things can be done with very simple mechanical switches, or with electrical switches called relays.

To make this point, a pupil has been given the logic circuit shown below. It is based on a relay.

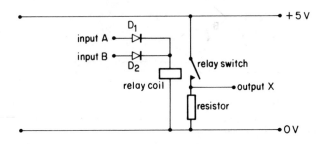

The pupil has been asked to connect the two inputs A and B to O or 5 volts in as many ways as possible, and to measure the voltage at X which is produced. She works systematically and carefully and produces the table shown below:

voltage at A	voltage at B	voltage at X
0 volts	0 volts	0 volts
0 volts	5 volts	5 volts
5 volts	0 volts	5 volts
5 volts	5 volts	5 volts

(a) Explain why this circuit is an OR gate. *(2)*
(b) Think of the case when A is connected to 5 V, and B is connected to 0 V. Why are the two diodes, D_1 and D_2 needed? *(2)*
(c) Why must there be a resistor in the circuit? *(1)*
(d) This circuit can only be used properly if the inputs to A and B change quite slowly. Explain why. *(2)*
(e) Draw a diagram to show how you would connect up a circuit using relays to make an AND gate. *(3)*

Total 10 marks

35 Heat energy

As part of an industrial process, a tank of oil has to be heated and then kept at a steady temperature. A graph of how the temperature changes with time is shown below:

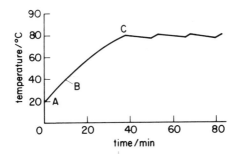

(a) Calculate the rate at which the temperature rises from A to B. *(2)*
(b) Explain why the graph from B to C does not show the same rate of rise of temperature. *(1)*
(c) What sort of device can be used to keep the temperature roughly constant at about 80°C? *(1)*
(d) How does the device mentioned in part (c) work? (There are several ways depending on how the temperature is measured and what sort of power supply is providing the heating energy. You should only describe *one*.) *(3)*
(e) If the tank contains 100 kg of a liquid which requires 2100 J to heat each kilogram by 1°C, calculate
   (i) the energy needed to heat all the liquid by 1°C
   (ii) the energy supplied to the liquid each second during the A-B section of the graph. *(2)*
(f) Why must the power of the heater be *greater* than your answer to e(ii)? *(1)*
(g) A piece of red hot metal is now plunged into the liquid (a process called quenching). Say what will happen to
   (i) the temperature of the liquid
   (ii) the heater supply to the tank
   (iii) the temperature of the metal. *(3)*
(h) If water (which requires 4200 J to heat each kilogram by 1°C) was used instead of the original liquid, how would your answers to (g)(i) and (g)(iii) change? *(2)*

Total 15 marks

36 Heat energy transfer and insulation

Heat energy can travel by conduction, convection and radiation. There are situations in the home where we want to cause the travel of heat energy. In an electric kettle we want all the water to get hot; the inside of an oven should be the same temperature all over; and a refrigerator should be equally cold all over.

However, there are also many situations where we *do not* want heat energy to travel. For this reason water pipes are lagged with foamed plastic; roof insulation made of 'glass wool' is used; another form of roof insulation uses shiny aluminium foil; the walls of refrigerators are made of foamed polyurethane sandwiched between two metal sheets; double glazing consists of two sheets of glass with a layer of air (or even better, a vacuum) trapped between.

Animals make use of the same principles to keep warm in winter. Polar bears have fur on the soles of their feet, birds fluff up their feathers and many animals grow longer coats in winter. Arctic seals have a thick layer of fat beneath their skin.

(a) Explain how *all* the water is heated by the element at the bottom of an electric kettle. (2)

(b) Explain why a simple oven is hotter at the top than the bottom, but the temperature is much more even in a 'fan assisted' oven. (2)

(c) Why are the cooling pipes in a refrigerator near the top? (2)

(d) The word 'foamed' is used in the passage above. Why are foamed materials good insulators? (2)

(e) Glass is a reasonably good insulator, but 'glass wool' is much better than solid glass. Why is this? (1)

(f) Metals, like aluminium, are very bad insulators (or very good conductors), so how can the foil mentioned in the passage help to keep heat energy inside a house? (1)

(g) Why is a vacuum better than air between the sheets of glass in double glazing? (2)

(h) Why do birds fluff up their feathers to keep warm? (2)

(i) What type of heat transfer is cut down by a seal's layer of fat? (1)

Total 15 marks

37 Expansion of gases

The apparatus shown below is used to measure the expansion of air. The air is trapped in a narrow tube of glass with thick walls, by a small quantity of mercury. The heating is done by placing the tube in water and heating the water. The expansion is found by measuring the length of the trapped air, and the temperature by placing a thermometer in the heating water. A careful student heats the water slowly, stirs well, and removes the heat so that the temperature shown on the thermometer remains steady for a while *before* each reading is taken. The results obtained are shown in the table. Experiments at very low temperatures show that the gas still behaves in the same way.

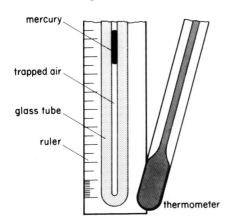

Results:

temperature /°C	length of air column/cm
20	7.5
44	8.1
68	8.7
100	9.5

(a) Draw a graph of length of air column against temperature, and state your conclusion. (5)

(b) If the tube was placed in a mixture of ice and water, what would be the length of the air column? (2)

(c) The temperature of the *air* is what is needed, but the temperature of the *water* is what is measured. Explain why the student carries out the heating in the way described. (3)

(d) If the air is cooled down as much as possible, at what temperature will the volume become almost zero? (3)

(e) Explain why the air expands when it is heated, in terms of the movement of its molecules. (2)

Total 15 marks

38 Measurement of energy for heating

A group of students was trying to measure how much energy was needed to heat different liquids, and set up the apparatus as shown. They made a serious mistake, but quickly realised this because the ammeter did not appear to register any current, and the heater did not warm up, even though the voltmeter gave a sensible reading. Fortunately, no damage was done.

Having put the circuit right, they went on to complete the experiment for water, and then repeated it with oil. Their results are shown below.

Readings when circuit correct:

Water:
mass	200 g
current	3.2 A
voltage	12 V

temperature/°C	20	25	30	35	39	42
time/min	0	2	4	6	8	10

Oil:
mass	200 g
current	3.2 A
voltage	12 V

temperature/°C	20	35	48	55	60	62
time/min	0	2	4	6	8	10

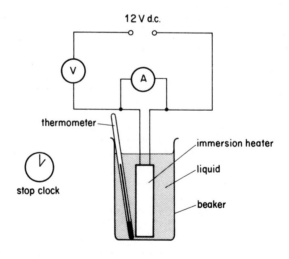

(a) (i) Should a good ammeter have a high or low electrical
       resistance?
    (ii) Should a good voltmeter have a high or low electrical
        resistance? *(2)*
(b) Why did the ammeter not show any noticeable current? *(1)*
(c) How should the circuit be corrected? *(2)*
(d) (i) Plot graphs of temperature (vertical axis) against time
       (horizontal axis) for water and for oil on the same axes.
    (ii) Explain why both graphs curve as the temperature rises.
    (iii) Why is the curve more obvious for oil? *(6)*
(e) The energy supplied by the heater in the first two minutes raises
    the temperature of the water by 5°C.
    (i) Calculate the power of the heater in watts.
    (ii) How much energy does the heater supply in each second?
    (iii) Calculate how much energy the heater supplied in 2 min.
    (iv) Calculate how much energy would be needed to raise the
        temperature of 1 kg of water by 5°C.
    (v) How much energy would be needed to heat 1 kg of water
       by only 1°C? *(5)*
(f) The figure calculated in (e) (v) is called the 'specific heat
    capacity' of water. The group look up the value in a book of
    physical constants, and find the value 4200 J/kg °C. One of the
    group says 'Typical, our experiments *never* give the right
    answers'. Explain how each of the following changes to the
    experiment would help to give a more accurate answer, so they
    need not feel so annoyed.
    (i) Surround the beaker with cotton wool.
    (ii) Put a lid on the beaker.
    (iii) Stir the liquid during the experiment. *(5)*
(g) Give one other way in which the experiment is not completely
    accurate, which is not helped by any of the changes in (f). *(2)*
(h) By comparing the graphs for water and oil during the first two
    minutes, calculate the 'correct' value for the specific heat
    capacity of the oil. *(2)*

Total 25 marks

39 Solar heating

The diagram shows part of a house hot water system. The 'solar panel' is placed on the outside of the roof of the house. The panel has a sheet of glass over it. The pipes inside the panel are painted dull black and are laid on a sheet of metal which is also black. There is a thick layer of insulation behind the metal sheet, and the pipes to the tank are also insulated. The hot water produced in the panel passes through a spiral of copper pipe inside the hot water tank. Because the sun does not always shine, the water in the tank can also be heated by using either of the electric immersion heaters. The hot water tank has a jacket of insulation around it.

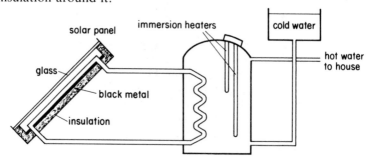

(a) Explain the reason for
  (i) the sheet of glass over the panel
  (ii) the black coating on the pipes and backing sheet
  (iii) the insulation behind the backing sheet. *(3)*
(b) Why is the backing sheet made of metal? *(1)*
(c) Suggest, with reasons, suitable materials for the insulation behind the panel, round the pipes to the hot water tank and the jacket round the tank itself. *(4)*
(d) Which way, and why, does the water circulate from the panel to the tank? *(2)*
(e) Why is the spiral in the tank made of copper? *(1)*
(f) The shorter of the two immersion heaters is used when only a small amount of hot water is needed. Why doesn't *all* the water in the tank get heated? *(2)*
(g) Explain why using the solar panel, rather than the immersion heaters, is an advantage to the householder and the environment. *(2)*

Total 15 marks

40 Thermostat

Many household heating devices, and industrial ones, need to keep a steady temperature. Examples are room heaters (or central heating), electric irons and coffee makers. One way of keeping this steady temperature is to switch the device off when the temperature gets too high, and switch it back on again when the temperature gets too low. The diagram shows the sort of switch (called a thermostat) that may be used to do this. The contacts are shown closed, so that when the device is switched on there is a complete circuit and current can flow. Heat from the device warms the bi-metal strip, which then bends and opens the contacts.

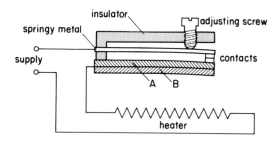

(a) A and B are both metals firmly fixed to each other. What is the difference between A and B? (2)
(b) Explain why the strip bends. (2)
(c) When the contacts have opened, what happens to the
    (i) current through the heater
    (ii) temperature of the heater
    (iii) shape of the bi-metal strip? (3)
(d) Turning the adjusting screw so that it moves downards increases the temperature at which the device works. Explain why this is so. (3)
(e) (Hard, optional) The contacts in the thermostat open and close very slowly. This causes sparking, which burns the contacts and causes radio interference. Making the contacts strongly magnetic, so that they attract each other, helps to cure this problem. Explain how this helps. (5)

Total 10 (15) marks

41 Furnace for melting metal

The furnace shown below in cross section uses electrical energy to heat the metal. A large coil, like a large electric fire, surrounds the crucible (the container for the molten metal). The furnace is surrounded by very good insulation to keep down the heating cost. In spite of this insulation, the temperatures reached are so high that some energy is still lost; in fact the furnace is 70% efficient.

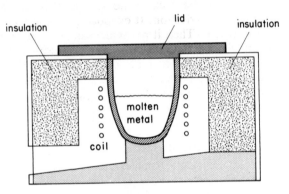

Information: specific heat capacity of iron = 460 J/kg °C
specific latent heat of iron = 275 000 J/kg
melting point of iron = 1540 °C

(a) How much of the energy supplied to the furnace is wasted? *(1)*
(b) Name three ways in which heat energy can escape from the furnace. *(3)*
(c) How do various parts of the insulation prevent these heat losses? *(3)*
(d) Some iron, mass 100 kg, is loaded into the furnace at 40 °C. Calculate
 (i) the temperature rise needed to reach the melting point
 (ii) the energy needed to heat the iron to its melting point
 (iii) the energy needed to melt the iron at its melting point
 (iv) the total energy needed to get the iron from 40 °C to molten iron
 (v) the energy supplied to the heating coils. *(8)*

Total 15 marks

42 Refrigerator

A modern refrigerator is designed using many ideas which relate to simple school physics. The sides are made of two thin sheets of steel directly bonded together with a thick layer of foamed polyurethane. The inside is moulded in smooth plastic. The door is held shut by a strip of flexible magnetic plastic inside the squashy, air-tight door seal. The 'fridge is cooled by pumping a liquid round through the pipes inside, then through pipes on the back of the 'fridge. This liquid must have a high latent heat of vaporisation. It evaporates in the pipes inside the 'fridge, absorbing heat. Then it re-condenses to liquid in the pipes on the back, giving out the heat. The pipes on the inside are concentrated near the top of the 'fridge. The pipes on the back run through a grid of copper strips, and the pipes and grid are painted matt black. When a refrigerator is fitted in a 'built-in' kitchen, a space must be left above the 'fridge to allow air to circulate.

(a) Why is *foamed* polyurethane used between the steel sheets? *(1)*

(b) Why is the inside moulding *smooth:* (The answer is more common sense than physics!) *(1)*

(c) Name a suitable powdered material to be mixed with the plastic strip during manufacture, so that it can be magnetised to make the door seal. *(1)*

(d) What does 'latent heat of vaporisation' mean? *(2)*

(e) Why are the cooling pipes inside the refrigerator near the top? *(2)*

(f) Why is the grid on the back painted black? (Hint: it has to get rid of heat energy.) *(1)*

(g) Which way does air circulate around the back of the 'fridge? Explain. *(2)*

Total 10 marks

43 Cooling investigation

A girl who was interested in mountain walking and camping decided to do a physics experiment to investigate different plastic cups. She wanted to know which of two cups would keep hot drinks hot for the longest time. The two cups were made of the same plastic, but were different shapes, and different thicknesses.

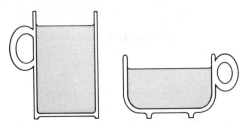

She poured the same amount of hot water into each cup, and recorded the temperature every minute. She then repeated the whole experiment, starting with hot water again, but this time she had a fan blowing cold air past both cups. A few of her results are shown below:

	tall cup (no fan)	short cup (no fan)	tall cup with fan	short cup with fan
time for temperature to fall from 80 °C to 40 °C	14 min	10 min	10 min	5 min

(a) How does the hot water lose energy to its surroundings
    (i) through the sides of the cups
    (ii) from the top of the cups? *(2)*

(b) (i) Which cup has the thinner sides?
    (ii) Which cup loses more energy through the sides? *(2)*

(c) (i) Which cup has water with more surface area?
    (ii) Which cup loses more energy from the water surface? *(2)*

(d) The answers to (b) (ii) and (c) (ii) show that one cup loses more energy through the sides, and the other one more from the top. By looking at the girl's actual results (without the fan), say which is the more important energy loss. *(2)*

(e) For this particular girl, with her interests, why is repeating the experiment with a fan a sensible thing to do? *(2)*

Total 10 marks

44 New clothing material

A new material has been invented for making warm clothing. The best sort of warm clothing up until now has had a windproof and waterproof cloth on the outside, some sort of insulation inside this, and then another layer of cloth on the inside. The insulation can be *down* (small, fluffy feathers) or some sort of fluffy fibre. This means that lots of air is trapped in the insulation. The new material uses fine metal coated plastic strips as the insulation. These trap air in just the same way as the normal insulation, but have the extra advantage that the metal coating is shiny.

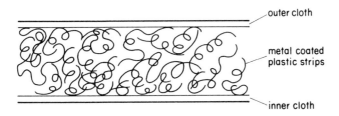

outer cloth

metal coated plastic strips

inner cloth

During tests on the new material it worked very well, except when the material was squashed. The metal coatings then touched each other, and there was a lot of heat loss. The problem has been overcome by making the strips especially springy and treating the edges so that the metal is much less likely to touch. The new material is as good as the best down filled materials, but is a lot cheaper.

(a) Name three ways in which heat can be lost from the body. *(3)*
(b) Why is air a poor conductor of heat? *(2)*
(c) The air trapped in the material described above cannot move around easily. What form of heat loss does this help to prevent? *(1)*
(d) The metal coated strips are shiny. What sort of heat loss does this cut down? *(2)*
(e) Why was there a lot of heat loss when the material was squashed, and the metal touched? *(2)*

Total 10 marks

45 Microwave oven

Think about baking potatoes in an ordinary oven. The heat has to heat up the outside of the potato. The heated outside passes some heat to the next layer in, then this passes some heat on to the next, and so on. By the time the inside is cooked, the outside may be blackened!

A microwave oven sends microwaves through the potato. They penetrate right through the potato *straight away*. This means that all of the potato is cooked at the same time. Naturally this makes the whole process much quicker, and uses much less energy.

The microwaves have to spread evenly through the oven. There is a spinning metal 'paddle' which reflects the waves in all directions around the oven. They bounce off the metal walls, so the whole space is filled with microwaves travelling in all directions. There may be some places where the waves combine to give a strong effect. There may be other places where they combine to give a weak effect. The food is often placed on a turntable to make sure the cooking is even.

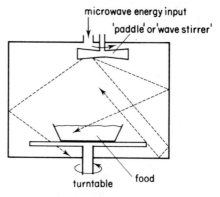

(a) Name the heat movement described in the first paragraph. *(1)*
(b) Why must all the potatoes baked in an ordinary oven at one time be roughly the same size? *(2)*
(c) Why should the food in a microwave oven *not* be placed inside a metal container? (Look in the passage to find out how microwaves behave when they hit metal.) *(2)*
(d) (i) How can microwaves combine to give a strong effect?
  (ii) How can they combine to give a weak effect? *(3)*
(e) How does the turntable help to give even cooking? *(2)*

Total 10 marks

46 Curved mirrors

When light is reflected off a mirror, the angle of incidence is equal to the angle of reflection. This is illustrated in fig. 1. If the mirror is a curved one, the light will be reflected in different directions at different parts of the mirror. In fig. 2, the beams of light which can be seen by reflection are coming from further apart than those in fig. 1. The mirror is a *convex* one.

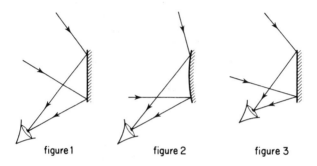

figure 1   figure 2   figure 3

(a) Why are convex mirrors used to spot shop lifters in large stores and supermarkets? *(3)*

(b) Give two other places or situations where mirrors like this are sometimes used. *(2)*

(c) Though they are useful, convex mirrors do have disadvantages. Look at the reflections on the back of a spoon and say what is odd about them. *(2)*

(d) The mirrors that a car driver uses to see behind are sometimes convex. Give one advantage and one disadvantage of using a convex mirror rather than a flat one. *(4)*

(e) One car maker uses a flat mirror on the driver's side, and a convex mirror on the other side of the car, furthest away from the driver. Figure 3 is similar to figure 1, but shows the eye nearer to the mirror. By comparing figure 1 and figure 3, explain why a flat mirror can be used for the mirror on the driver's side. *(4)*

Total 15 marks

47 Total internal reflection

If a beam of light travelling through glass (or plastic) comes to the edge
of the material at a large enough angle, total internal reflection takes
place. If the angle is too small, light will escape. This is shown in
figure 1. The principle is now being used in 'optical fibres' for telephone
signals, various medical uses and for many other purposes. The start of
an optical fibre is shown in figure 2.

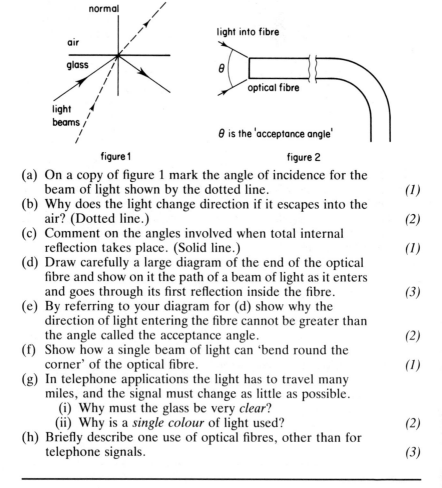

figure 1   figure 2

(a) On a copy of figure 1 mark the angle of incidence for the
    beam of light shown by the dotted line.   *(1)*
(b) Why does the light change direction if it escapes into the
    air? (Dotted line.)   *(2)*
(c) Comment on the angles involved when total internal
    reflection takes place. (Solid line.)   *(1)*
(d) Draw carefully a large diagram of the end of the optical
    fibre and show on it the path of a beam of light as it enters
    and goes through its first reflection inside the fibre.   *(3)*
(e) By referring to your diagram for (d) show why the
    direction of light entering the fibre cannot be greater than
    the angle called the acceptance angle.   *(2)*
(f) Show how a single beam of light can 'bend round the
    corner' of the optical fibre.   *(1)*
(g) In telephone applications the light has to travel many
    miles, and the signal must change as little as possible.
       (i) Why must the glass be very *clear*?
      (ii) Why is a *single colour* of light used?   *(2)*
(h) Briefly describe one use of optical fibres, other than for
    telephone signals.   *(3)*

Total 15 marks

48 Pavement lights

In towns you may have seen square blocks of glass set into the pavement. They are there to let light into the basements of buildings. Figure 1 shows what they look like from above, and figure 2 shows a side view of one block. Light bends as it goes through the block, so light from outside is sent downwards into the basement.

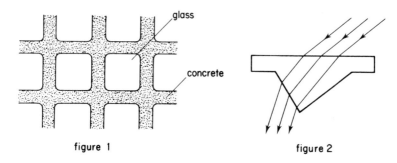

figure 1   figure 2

Different weather conditions may produce different problems; when it is wet the blocks may be dangerous to walk on, and on hot days the block and the concrete they are set in may expand by different amounts.
(a) Why does the light bend as it enters the glass block?   (2)
(b) Why is the building on the left of the block shown in
     figure 2?   (1)
(c) Why could the blocks be dangerous in wet weather?   (1)
(d) What might happen in very hot weather if
      (i) the concrete expanded more than the glass blocks?
     (ii) the glass blocks expanded more than the concrete?
          (Fortunately they both expand by almost the same
          amount.)   (4)
(e) Pavement lights are very rarely used now when new
     buildings are put up. Why do you think there is less
     demand for natural light in basements these days?   (2)

Total 10 marks

49 Light and police work

After a burglary, if a window has been broken, there will be tiny particles of glass. Some of these will be found at the scene of the crime, and some may be caught in the burglar's clothing. If the police can prove that these particles are identical, they will have a strong case. A method of doing this is to suspend the particles in a special liquid. Light of a single colour is shone through the liquid and the particles viewed through a microscope. The temperature of the liquid is then slowly altered. This alters the speed of light through the liquid (i.e. it alters the refractive index). At one particular temperature the particles of glass disappear. If this happens at the same temperature for both sets of glass particles, they probably came from the same broken pane of glass.

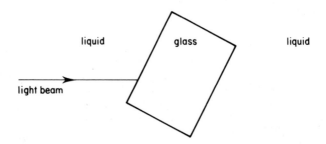

(a) Complete a copy of the diagram to show how light bends when it travels from the liquid to the glass, and back to liquid,
  (i) if the light slows down in the glass
  (ii) if the light speeds up in the glass. *(4)*
(b) Explain why the bending in (a) (i) and (ii) takes place, in terms of light as a wave motion. *(2)*
(c) Draw a diagram to show how different colours of light are bent by different amounts. *(2)*
(d) Why is the light used in the police work a *single* colour? *(2)*
(e) Under what *two* circumstances can light moving from liquid to glass pass straight on without bending? *(2)*
(f) For tiny, irregular particles of glass, only one of (e) is possible. Which one? *(1)*
(g) When do the particles of glass disappear in the police method? *(2)*

Total 15 marks

50 Sunshine detector

Weather forecasting is very difficult to get right, because the weather is very complicated. Part of the job is to collect accurate information about what the weather is actually like, so that future forecasts can be improved. Recording the hours of sunshine can be done using the device described below.

The diagram shows a spherical glass lens with a focal length of 30 cm. It is held in front of a piece of paper, which is replaced each day. The paper is bent round so that each part of it is the same distance from the lens. The paper blackens when it gets hot, but does not catch fire. The paper is printed with the time of day along it. A typical record for part of a day is also shown.

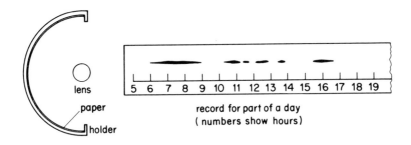

record for part of a day
(numbers show hours)

(a) How far should the lens be held from the paper to get the maximum blackening effect? *(1)*

(b) If the lens got broken, and only the replacement was a less powerful lens (i.e. one with a longer focal length), how should the lens holder be re-positioned? *(2)*

(c) Draw a diagram to show rays of sunlight passing through the lens and reaching the paper. *(2)*

(d) Why is the paper bent round in the way described in the passage? *(1)*

(e) How many hours of sunlight were there on the day shown? *(1)*

(f) Why is the paper strip not evenly blackened, even when the sun was shining? *(1)*

(g) Describe what the weather might have been like between 10.00 and 14.00 hrs. *(2)*

Total 10 marks

53

51 Industrial uses of fibre optics

The diagrams below show three uses of fibre optics.

In the first one, a light is needed on a particular part of a vibrating machine. If the bulb is placed on the machine the vibration will shorten its life.

The second use is to sense the level of liquid in a tank. If light enters the sensor nothing happens, but if there is no light a warning is given.

Finally, a beam of light given out by one bundle of optical fibres on one side of a conveyor belt is collected by another bundle on the other side. Every time an object on the belt breaks the beam an automatic counter operates.

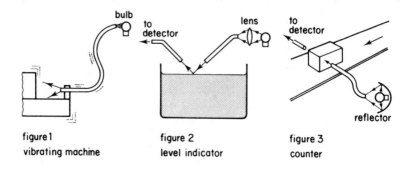

figure 1
vibrating machine

figure 2
level indicator

figure 3
counter

(a) How can the light travel along a bundle of fibres even when it goes round corners? (3)
(b) Draw diagrams to show why no light enters the sensor if the liquid level is
  (i) too low
  (ii) too high. (2)
(c) Explain why the warning may sound, even if the liquid level is correct, if the surface of the liquid is rough. (2)
(d) (i) Explain why using a lens (figure 2) is better than just a bulb (figure 1).
  (ii) Why is the reflector (figure 3) even better? (3)

Total 10 marks

52 Prism spectacles

Some hospital patients have to lie flat on their backs for long periods of time. Depending on how much they can move their heads, they may have a very boring view of the ceiling!

An obvious solution is to hang a mirror above their heads at about 45°. However, unless the mirror is very large, they may only be able to see a small area of the room. Also, everything looks upside down!

Another solution is to wear prism spectacles. The light from objects in the room changes direction, and so can enter the patient's eyes. If some head movement is possible, the spectacles also move, so vision is not limited. More of the room can be seen if the prisms can rotate in the spectacle frame. Better still, the picture seen is the right way up. However, there is one snag: objects may appear to have coloured edges.

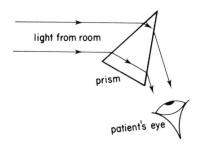

(a) Use a sketch diagram of rays of light bouncing off a mirror to explain why the picture is upside-down. *(2)*

(b) Show how a convex mirror could give a wider view. (You may have seen a mirror like this in a large shop, or at the top of the stairs on a double-deck bus.) *(2)*

(c) What causes the light to bend when it passes through the prism? (See diagram.) *(2)*

(d) The prisms can be made of glass or plastic. Give one advantage of using glass, and one of using plastic. (They do not have to be *optical* advantages.) *(2)*

(e) On a copy of the prism part of the diagram show how a beam of white light is split into a spectrum. *(2)*

Total 10 marks

53 Faulty vision

The diagram shows a cross section through a human eye. The almost parallel rays of light are from an object a long way away. For a person with normal vision, these rays would be accurately focussed on the retina. However, in the eye shown they are focussed in front of the retina. By the time the light reaches the retina it has spread out, and the person sees a blur. This is called 'short sight'.

Short sight can be corrected by spreading the light out using spectacle lenses or contact lenses. However some surgical methods have been tried in severe cases. One method is to cut a small pice out of the side of the eyeball, and then stitch it together again. Another method is to adapt a standard operation called a 'corneal graft'. While the patient is unconscious the cornea is cut off, frozen solid, its front surface re-shaped, unfrozen and stitched back on again! Yet another possibility, but only for children, is to wear contact lenses which deliberately do not fit. As the child grows, the eyeball is encouraged to change shape to match the contact lens.

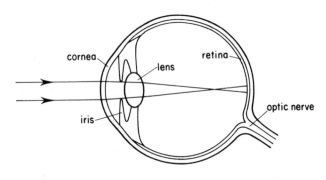

(a) To correct short sight, what sort of lenses should be used? *(2)*

(b) (i) How does cutting a piece out of the eyeball change its size?

    (ii) How does this correct for short sight? *(4)*

(c) In the corneal graft type of operation, should the cornea be made more sharply curved or flatter? *(2)*

(d) Sketch the cornea and the shape of the contact lens that will produce the correction described in the last method. *(2)*

Total 10 marks

54 Different eyes in nature

The diagrams on the opposite page and on this page show cross-sections through a human eye and the eye of a nautilus, a type of shell-fish. The human eye has many features which are like a camera. The eye of the nautilus is much simpler, very much like a pinhole camera.

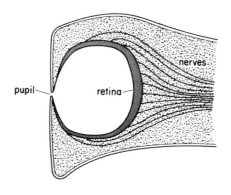

(a) How does the human eye control the amount of light which can get in? *(1)*

(b) How must the shape of the lens in the human eye change to give a sharp, focussed image for light from
   (i) a distant object
   (ii) a close object? *(2)*

(c) A camera has a glass lens, so its shape cannot be changed. How is the focussing done in this case? *(2)*

(d) The nautilus has no method of focussing. Explain how it can see distant objects and close objects clearly. *(2)*

(e) The eye of the nautilus has the disadvantage that it lets in much less light than the human eye. What would be the bad effect of simply opening up the hole to let more light in? *(1)*

(f) The human eye seems to be much superior to the nautilus eye. Why doesn't a biologically simple creature have an eye like a human? *(2)*

Total 10 marks

55 Traffic lights

Traffic lights have a bright bulb in each of the three coloured lamps. So that the lights look even brighter, there is a mirror behind each bulb. In most situations this works well. However, if bright sunlight is shining straight into the light, the mirror sends the light back out. This makes it hard to tell which light is on. Fibre optics can be used to solve the problem. A much smaller mirror is put behind the bulb, and the light spread out using the fibres as shown.

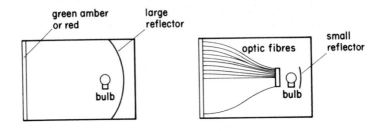

(a) Draw a diagram showing how the light from the bulb is reflected off the large mirror. (2)

(b) Draw a diagram showing how the light from the sun is reflected back out from the large mirror. (2)

(c) Why does the small mirror reflect less sunlight? (1)

(d) What sort of reflection is happening inside the fibres? (1)

(e) Using a diagram to help, explain how the light from the small mirror travels along one of the fibres, even though the fibre bends round corners. (2)

(f) Each optic fibre is surrounded by a plastic tube to protect it. What optical property must the plastic have, so that the reflection referred to in (d) can happen? (2)

Total 10 marks

56 Expansion measured using interference

Some engineering materials are used at very high temperatures. (Think of jet engines, nuclear reactors, space vehicles re-entering the atmosphere etc.) It is important to know how much these materials expand when heated. One way of measuring the expansion is to use the inteference of light waves. The diagram shows a beam of light from a laser. Some of this light is reflected off the top piece of sapphire, and some off the lower piece of sapphire. When these two parts of the beam re-combine, they have travelled different distances, and so the light waves could add together or cancel. When the material between the two pieces of sapphire expands, the extra distance travelled by the bottom light beam changes, and so the interference pattern changes. This is used to measure the expansion.

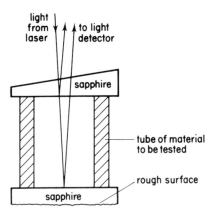

(a) What must the light waves be like to
  (i) add together
  (ii) cancel? *(3)*
(b) Light reflected off the top surface of the top piece of sapphire could complicate things. How does the fact that this surface is sloping avoid this complication? *(2)*
(c) Why is the bottom of the lower piece of sapphire roughened? *(2)*
(d) If the wavelength of light being used is 630 nm, and the inteferrence pattern changes from bright to dark and back to bright 20 times, calculate the expansion of the material.
  (1 nm = 1 thousand millionth of a metre.) *(3)*

Total 10 marks

57 School journey

A school party goes skiing in Austria. The approximate times and distances for the coach journey are shown in the table below.

	time	distance/km (from start)
London	10.00	0
Folkestone (arrival)	11.40	100
Folkestone (departure)	12.30	100
Boulogne	14.15	150
Austria	10.15 (next day)	1150

(a) (i) How many minutes did it take to get from London to Folkestone?

(ii) What was the average speed (in km/h) for this part of the journey? *(3)*

(b) (i) How far is it from Folkestone to Boulogne?

(ii) How long did it take to cross from Folkestone to Boulogne?

(iii) What was the speed of the crossing? *(4)*

The coach drivers have a much more detailed record of the journey. The coach is fitted with a 'tachograph'. This contains a paper disc which rotates slowly. A pen draws a line on the paper as it rotates. The faster the coach goes, the further the pen moves out from the centre of the disc. The disc is shown in the diagram below. (Only the first 6 hours are marked in.) The scale marked out from the centre is in kilometres per hour, and the lines from the centre are at 1 hour intervals. A graph obtained from another (magnified) part of the disc is also shown.

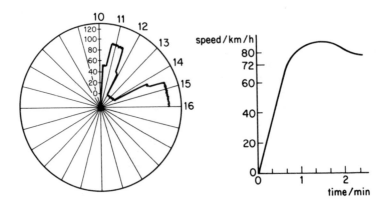

(c) There are many variations in speed in the first two hours, but the journey can be roughly divided into three parts. Estimate the speed for each of these three parts, and say why you think the journey was like this. (Think about road conditions and traffic.) *(6)*

(d) Estimate the maximum speed reached in the section shown on the disc. *(2)*

(e) Use the table at the beginning to answer this question.
    (i) How far is it from Boulogne to the ski resort in Austria?
    (ii) How long did it take to travel this distance?
    (iii) Calculate the average speed for this part of the journey. *(3)*

(f) Use information from the graph to answer this question.
    (i) Convert 72 km/h to a speed in m/s.
    (ii) How many *seconds* did it take the coach to reach a speed of 72 km/h?
    (iii) Calculate the accceleration of the coach, in m/s per s.
    (iv) If the coach has a mass of 16 tonnes (16 000 kg), calculate the force used to cause this acceleration. *(7)*

_____

Total 25 marks

58 Launching a space ship

Most of the mass of a space ship on the launching pad is fuel. When this fuel is burned, the space ship accelerates upwards, slowly at first. As the ship gets higher the acceleration increases. Eventually a particular ship, of final mass 1000 kg, reaches a speed of 3000 m/s in orbit round the earth, well above the atmosphere. The engines are then turned off, but the ship keeps moving at a steady speed. Later the engines are turned on again, and produce a thrust of 2000 N for 100 s. The space ship now heads towards the moon with the engines turned off again. The speed is now found to be decreasing (though the astronauts on board have no sensation of any speed). When the ship gets much nearer to the moon the speed starts to increase again. After a successful landing on the moon the space ship must still have enough fuel to return to earth.

(a) The fuel contains a large amount of stored energy. State *three* kinds of energy that this is converted to as it accelerates the space ship. *(3)*

(b) Give two reasons for the increase in acceleration as the ship climbs. *(2)*

(c) When the ship has reached orbit, calculate
    (i) its momentum
    (ii) its kinetic energy. *(2)*

(d) Explain why the space ship continues at a steady speed, even though its engines are turned off. *(2)*

(e) When the engines are turned on again, calculate
    (i) the acceleration produced
    (ii) the increase in speed produced (assuming the acceleration is in the same direction as the original speed). *(2)*

(f) On the trip to the moon, why does the speed of the space ship decrease and then increase again, without the engines being used? *(2)*

(g) Give two reasons why the ship needs much less fuel to lift off from the moon than it did from the earth. *(2)*

Total 15 marks

59 Loading a lorry

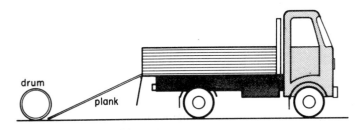

A lorry driver can load oil drums into the back of the lorry by pushing them up a sloping plank, or by lifting them directly. Each drum has a mass of 80 kg, the plank is 3 m long, and the back of the lorry is 0.8 m above the ground.

(a) How much force would be needed to lift a drum into the lorry directly, *without* using the plank? (Take the earth's gravitational field as 10 N/kg). *(1)*

(b) How much energy would be converted in lifting the drum into the lorry *without* the plank: *(2)*

(c) The force needed to push a drum up the plank is 300 N. Why is this less than the answer to part (a)? *(2)*

(d) How much energy is converted in rolling a drum from the bottom of the plank to the top? *(2)*

(e) Your answer to part (d) should be greater than your answer to part (b), so using the plank means that the driver has to convert *more* energy. Why does he use the plank? *(2)*

(f) When the lorry is loaded, the driver drives off. List the energy changes that take place in moving the lorry. *(4)*

(g) The driver has to stop at the factory gates. What happens to the kinetic (movement) energy of the lorry? *(2)*

Total 15 marks

60 The physics of a ski lift

A ski lift consists of a loop of cable which passes round a large pulley wheel at the top of a slope, and round another one at the bottom of the slope. It also passes over or under other pulleys attached to pylons which keep the cable at roughly the right height above the slope. Each skier takes hold of a device attached to the cable. The large pulley wheel at the bottom of the slope is driven round by an electric motor. Each skier is then literally dragged up the slope, unless they fall off!

A particular lift has a length of 400 m, but only rises a vertical height of 200 m. When it is running normally there are 25 skiers on the lift at any time. The speed of the cable is 1.6 m/s. The operator can stop the lift if someone falls off and does not get out of the way of the next skier in time. When the way is clear again the operator re-starts the lift.

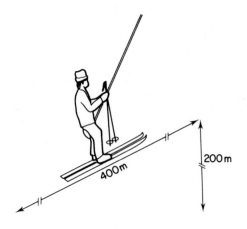

(a) How far apart are the skiers? *(1)*
(b) What is the time interval between one skier arriving at the top of the slope and the next? *(2)*
(c) Each skier is travelling at 1.6 m/s along the slope. How far does each skier rise *vertically*, on average, in one second? *(2)*
(d) Why in question (c) does it say 'on average'? (Think about the exact shape of the slope.) *(2)*

(e) Calculate the increase in potential energy of one skier in one second, assuming a mass of 75 kg and taking the earth's gravitational field strength as 10 N/kg. *(2)*

(f) Now calculate the increase in potential energy in one second for all the skiers on the lift. *(2)*

(g) What is the power of the electric motor needed *just* to provide the energy calculated in (f)? *(2)*

(h) Some of the cable and devices that the skiers hand on to are also being lifted. Why would it be wrong to include them in the calculation for (g)? *(2)*

(i) If the lift is stopped, the motor must also be powerful enough to get it started again. Calculate how much kinetic energy the motor would have to give to
    (i) each of the skiers, and
    (ii) all the skiers
to get them moving at 1.6 m/s. *(3)*

(j) Give two other reasons why the motor will have to supply more energy than the calculation in (g) *(2)*

Total 20 marks

61 Piledriver

When a new road is being built along a hillside there is a danger that the foundations may slide down the hill. One way to avoid this is to drive 'piles' deep into the ground. A pile is a girder-like piece of steel. Many of these driven in side by side give a firm base on which to build the road. A piledriver is the machine used to drive the piles into the ground. One type uses a diesel engine to run an air compressor. When the air has built up enough pressure it makes a mass of 500 kg on top of the pile jump up and then fall again. The mass jumps up with a speed of 2 m/s. The pile gets a push into the ground when the mass jumps up, and again when it lands back on the pile. The pile only goes 5 cm into a particular piece of ground each time the mass lands, but the process can be continued as long as necessary.

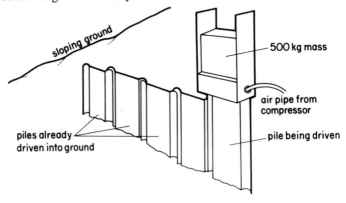

(a) How much force does the mass exert on the pile when it is *stationary*? (Earth's gravitational field strength = 10 N/kg). (2)
(b) Calculate (i) the momentum and (ii) the kinetic energy of the 500 kg mass moving at 2 m/s. (2)
(c) How much potential energy does the 500 kg mass have at its highest point, before it falls back towards the pile? (2)
(d) Use your answer to (c) to calculate how high the mass rises. (2)
(e) How fast is the mass moving at the bottom of its fall? (2)
(f) On hitting the pile, the mass slows down and stops. Calculate
    (i) its average speed during this slowing down
    (ii) the time taken to stop, from the moment of impact. (3)
(g) Calculate how much force the 500 kg mass exerts on the pile, because of the impact. (2)

Total 15 marks

62 Safety in rock climbing

In 'free' rock climbing (as opposed to 'artificial' climbing) a rope is used only as a safety measure. The leader of the climb climbs the first 'pitch' dragging the rope behind her. She then ties herself onto the rock face (called 'belaying') and takes in the slack of rope as the second person climbs. If the second person falls off, the rope stops him falling very far. When the second reaches the belay, he ties himself on and the leader climbs the next pitch. If the leader falls off she will fall quite a distance before the rope becomes tight and stops her falling any further. It is important that the rope has a lot of 'give' in it so that the falling climber is stopped gradually, rather than suddenly. Nylon is a material commonly used for climbing ropes, because it is strong but stretchy. Where necessary in the following questions, take the mass of the climber to be 70 kg and the earth's gravitational field strength to be 10 N/kg.

(a) The leader is 2.5 m above the belaying point when she falls.
   (i) How far does she fall before the rope becomes tight?
   (ii) How much potential energy does she lose in falling this distance?
   (iii) Calculate her speed of fall when the rope becomes tight.
   (iv) How long does it take the climber to fall this distance? *(7)*

(b) During the slowing down, the forces acting on the climber are complicated. There are two forces acting at once, and one of them is changing all the time.
   (i) What two forces are acting?
   (ii) Which of these is changing? *(3)*

In the following questions, ignore the weight of the climber. This means your answers will only be approximate.

(c) The climber is slowed down by the rope, over a distance of 1 m, from the speed you calculated in (a) (iii) to zero.
   (i) What is her average speed during this slowing down?
   (ii) How long does the slowing down take?
   (iii) How much momentum has the climber lost?
   (iv) What is the average force exerted by the rope? *(5)*

Total 15 marks

63 Accelerometer

An accelerometer measures acceleration. The type described in this passage is self-contained; it does not need to measure speed relative to the ground. This makes it suitable for use in rockets.

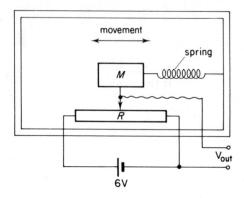

The accelerometer is firmly fixed in the rocket. The mass M (0.1 kg) is held in guides so that it can easily slide along the direction shown. If the rocket is stationary, the spring holds M in position in the middle of the casing. Now imagine that the rocket (and therefore the casing) accelerates from left to right. The inertia of the mass means that it tends to be left behind. The spring becomes stretched and so pulls on M accelerating it. The amount of force needed to stretch the spring by different amounts is shown in the graph.

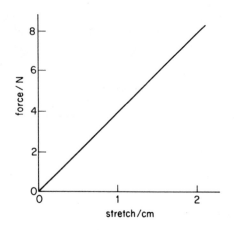

When the acceleration of *M* matches the acceleration of the rocket, the spring remains stretched by a fixed amount. The amount of stretch of the spring is converted into an electrical signal by the slider on the resistor.

(a) How much force on the spring will stretch it by 2 cm? *(1)*

(b) How much does the spring stretch if a force of 1 N is used? *(2)*

(c) If the force remains steady at the values in (a) and (b), calculate the acceleration of the mass in each case. *(2)*

(d) If the maximum movement of the spring is 3 cm before *M* reaches the end of the guides, calculate the maximum force that the spring can exert. What assumption must be made to do this calculation? *(3)*

(e) Calculate the maximum acceleration that can be measured by this accelerometer. *(2)*

(f) When the rocket has finished accelerating, and is travelling at a steady speed, where is the mass *M*? Explain your answer. *(2)*

(g) What is the output voltage that corresponds to
    (i) no acceleration
    (ii) maximum acceleration
    (iii) maximum *deceleration*? *(3)*

Total 15 marks

64 Impact testing of concrete pipes

Concrete pipes for drainage are laid in trenches dug in the ground. The pipes are covered using fairly fine, soft material before the trench is re-filled. A lot of time and money could be saved if the trenches could be re-filled by just bulldozing earth and rocks back into the trench. However, this may break the pipes. The method shown below is used to test the pipes by dropping a heavy weight on them.

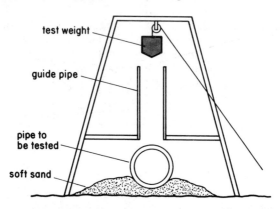

(a) The earth's gravitational field is 10 N/kg approximately. If the mass used is 20 kg and it is raised 2 m above the pipe,
  (i) calculate its greatest potential energy
  (ii) what is its kinetic energy just before striking the pipe
  (iii) how fast is the weight falling when it hits the pipe?
  (iv) How would your answers to (i), (ii) and (iii) be different if the mass was only 10 kg? *(6)*
(b) Explain why different shapes of weight would have different effects when they strike the pipe. *(2)*
(c) Why is the pipe to be tested resting on a bed of soft sand? *(2)*
(d) The flexible anti-corrosion coating on the pipes may help to protect them from impact damage. In the same test as (a) the weight is stopped in 1/100 s by an un-coated pipe, but in 1/10 s by a coated pipe. Calculate
  (i) the momentum of the weight just before it hits the pipe
  (ii) the force on the un-coated pipe
  (iii) the force on the coated pipe. *(5)*

Total 15 marks

65 Towing at sea

When ships have engine trouble they may have to be towed to port. Large structures like oil drilling platforms have to be towed into place. These jobs can be difficult, because large forces are needed and the sea may be rough. The towing may be done with several tugs pulling in slightly different directions. They may use very long cables. The details of a particular towing job are shown below:

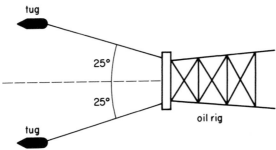

Mass of oil rig; 2000 tonnes.　　　　Max. force in cable; 330 000 N.
Combined pull from cables; 200 000 N. Max. stretch of cable; 0.45%.

(a) The oil rig is stationary to start with. The combined force of 200 000 N is applied (very carefully!). Calculate the acceleration of the oil rig. (1 tonne = 1000 kg) *(2)*

(b) If the two tugs apply the same force, in which direction does the oil rig move? *(2)*

(c) Explain (by calculation if you can) why the force in each cable is approximately 110 000 N. *(3)*

(d) As the oil rig picks up speed, the acceleration *decreases*, although the tugs keep the same forces in the cables. Explain. *(2)*

(e) Eventually the oil rig is being pulled along at a steady speed.
　　(i) What is the net (total) force acting on the oil rig?
　　(ii) How big is the drag (friction) force from the water? *(2)*

(f) Calculate the stretch of one of the cables when it has a force of 110 000 N in it, if it is
　　(i) 100 m long
　　(ii) 500 m long. *(2)*

(g) Why are towing cables usually quite long? (Look at your answers to (f), and think of the effect of a rough sea.) *(2)*

Total 15 marks

66 Forces in cables

The picture below shows the sort of cables used to get electricity to isolated places. The cables cannot always go in straight lines because of the lie of the land. Pole A is at a corner, but B is on a straight stretch. The weight of the cables pulls on the poles.

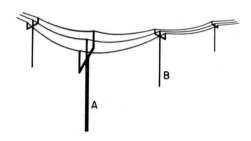

In this situation the people installing the cables may want to measure the tension in the cables (with the electricity turned off, of course!). An engineer testing the safety of a lift might also want to measure the tension in the cables supporting the lift. A device has been developed which clips onto the cable to measure the tension in it. The principle is shown in the diagram below. The rollers X and Y are fixed. A force F is needed to push roller Z so that it puts a link in the cable. By measuring F, the tension in the cable can be found.

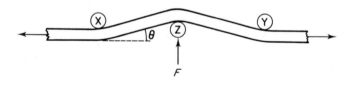

(a) In which direction do the tensions in the cables pull on pole B? *(1)*
(b) In which direction do the tensions in the cables pull on pole A? *(2)*
(c) Pole A can be steadied by using a wire attached below the cross-piece and firmly fixed in the ground. On which side of A should this wire be fixed? *(1)*
(d) If the cables are put up on a hot day in summer, explain what will happen to the tension in the cables on a very cold day in winter. *(2)*
(e) How does the force F depend on
  (i) the tension in the cable?
  (ii) the angle θ? *(2)*
The next question is only for those who can add and resolve vectors.
(f) If the tension in the cable is 1000 N and the angle θ is 20°, calculate
  (i) the tension in the part of the cable from X to Z
  (ii) the force F. *(2)*

Total 10 marks

67 Wave energy

Because of problems with energy sources used at present, there has been a lot of research into 'renewable' energy sources. One of these is to use the energy supplied by sea waves. In one proposed method, very large, flexible air bags are floated just below the surface of the sea. As the waves arrive, they compress the air in the bags at different times. Air is driven through connecting tubes from one of the compressed bags into one which is not being compressed. The moving air drives turbines round. The turbines in turn drive generators which produce electricity. At a later time the air will be driven back again, and more electricity can be generated.

Information about the device:

one unit			
length	275 m	number of units	320
depth	15 m	cost to build	£3000 million
width	13 m	cost of energy per kWh	5 p
max. power rating	10 MW		

Information about Atlantic waves (typical values):

frequency	0.1 Hz	power per m length	40 kW
wavelength	150 m	of wave 2 m high	

(a) Calculate the speed of a typical Atlantic wave. *(2)*
(b) What is the time interval between one wave and the next? *(2)*
(c) If two bags are 30 m apart, in the direction in which the waves are travelling, how long does it take a wave to get from one bag to the other? *(2)*
(d) The table gives the power of a wave per metre of its length.
    (i) How much power is this over the length of one unit?
    (ii) Is this more or less than the maximum power rating?
    (iii) Do you think the device can convert *all* the power available in a wave?
    (iv) Is the maximum power rating of 10 MW high enough? *(4)*
(e) Over what length of coastline will the 320 units stretch? *(1)*
(f) What effect would the device have on the waves passing it and reaching the shore? *(2)*
(g) Give two possible disadvantages of the scheme. *(2)*

Total 15 marks

68 Severn barrier

Coal fired power stations produce electrical energy by burning coal, which cannot be renewed. Nuclear fuels will last very much longer, but produce dangerous waste. A tidal barrier across the mouth of the river Severn has been proposed as an alternative, renewable, source of energy. As the tide rises, the water would be allowed to flow through open gates in the barrier. At high tide the gates would be closed. The trapped water would be allowed to run back out through large tubes containing turbines. As the turbines were forced round they would drive generators which would produce electrical energy.

The table shows the relative costs of three types of power station.

	coal-fired power station	nuclear power station	Severn barrier
total building and running cost per mega joule of energy delivered	1.3 p	0.6 p	1.0 p

(a) What will eventually happen to coal if we continue to burn it? *(1)*
(b) What other disadvantage is there to burning coal? *(2)*
(c) Nuclear energy is the cheapest form listed in the table, but there is a lot of public concern about it.
   (i) What could happen if there was a serious accident at a nuclear power station?
   (ii) Why is the waste from a nuclear power station dangerous? *(2)*
(d) (i) What form of energy will the water have when it is trapped behind a barrier at high tide?
   (ii) Calculate this energy for 1 kg of water, at a height of 5 m above the turbines. (Take the earth's gravitational field as 10 N/kg)
   (iii) To what form is this energy converted as the water flows through the turbines? *(3)*
(e) The other types of power station have serious environmental disadvantages. Give one effect that a tidal barrier would have on its surroundings. *(2)*

Total 10 marks

69 Safety in cars

Car makers build in many safety features to their cars. Some advertise these features very much, so that many people will have heard of 'crumple zones' and 'rigid steel safety cage'. Seat belts have also led to increased safety.

The crumple zones in a car are the engine compartment and the boot. These are designed so that they crumple or collapse in a serious collision. This means that the impact is spread over a longer time. The safety cage is a cage of strong steel bars around the passenger space, hidden in the body of the car.

A motoring correspondent on a radio programme once said that everyone would be safer if all cars were very flimsy, and had a large spike in the middle of the steering wheel, pointing at the driver!

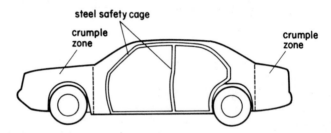

(a) A car of mass 900 kg is travelling at 10 m/s when it skids off the road and into a concrete wall. If the car does not have crumple zones it is stopped in 0.5 s. If it does have crumple zones the slowing down process takes much longer, 4.5 s.
   (i) What is the momentum of the car before it hits the wall?
   (ii) What is the momentum after the collision?
   (iii) What is the change of momentum?
   (iv) How big is the force stopping the car if it does not have crumple zones?
   (v) How big is the force if it does have crumple zones?   (5)
(b) In the crash in (a) the driver is likely to be injured. Say how you think the driver's body will react during the crash, with and without a seat belt.   (3)
(c) Why does the cage make the passengers safer?   (1)
(d) What serious safety point was the motoring correspondent trying to make by his ridiculous comment?   (1)

Total 10 marks

70 Acoustic strain gauge

A wire stretched between two points can vibrate at a particular frequency. This is the basis of stringed musical instruments like guitars and violins. If the string is pulled tighter, it will vibrate at a higher frequency; if it is relaxed it will vibrate at a lower frequency. This idea is also used in the acoustic strain gauge, shown below.

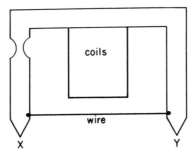

The device is clamped to a piece of metal which may stretch in use. (Examples are girders in bridges and the metal shaft driving the propellor of a large ship.) It is clamped so tightly that the hardened points X and Y dig into the metal. If the metal stretches, the points X and Y move apart, changing the tension of the wire. This alters the frequency of vibration of the wire. The wire is made to vibrate by energy from one of the coils, and the actual vibration is detected by the other coil. Any change in the frequency of vibration can be used to work out how much strain there is in the metal to which the device is clamped.

(a) What does frequency of vibration mean?   *(2)*

(b)   (i) If the hardened points of the strain gauge are pulled further apart, does the tension in the wire increase or decrease?

     (ii) Does this increase or decrease the frequency of vibration? *(2)*

(c) Why is there a narrow part in the frame holding the wire?   *(2)*

(d)   (i) The coil which makes the wire vibrate has an electric current passed through it. What sort of field does it produce?

     (ii) The wire has to be affected by this field. From what material should it be made?   *(2)*

(e) If the wire was replaced by a thicker one, how would this alter the vibration of the wire?   *(2)*

Total 10 marks

71 Mass of an astronaut

When astronauts are in a satellite orbiting the earth they are in free-fall, and feel 'weightless'. Ordinary scales cannot be used to measure their body mass. However, it is important to measure their mass, to keep a check on their health. One way to do this is to use a couch held in place by springs. If the astronaut pushes to one side and then lets go, the couch and astronaut will oscillate from side to side. The frequency of oscillation depends on the mass of the astronaut, and on the strength of the springs.

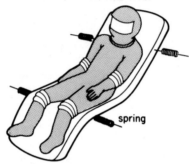

spring

(a) Kinetic energy (meaning movement energy) and potential energy (meaning stored energy) are involved in the oscillation.
   (i) Where is the kinetic energy at its maximum?
   (ii) Where is the kinetic energy at its minimum?
   (iii) Where is the potential energy at its maximum?
   (iv) Where is the potential energy at its minimum? *(4)*
(b) Question (a) reminds you that potential energy is stored energy. *Where* is the energy stored? *(1)*
(c) The speed of the couch and astronaut changes during each oscillation. What happens to the kinetic energy if the speed doubles? *(1)*
(d) When the astronaut and couch are at one side of the oscillation, they are pulled back to the middle by the springs. How will the speed at which this happens change if
   (i) the springs are made weaker
   (ii) the mass of the astronaut decreases? *(2)*
(e) Does the frequency increase or decrease in (d) (i) and (ii)? *(2)*

Total 10 marks

72 Checking a baby's breathing

Some babies are born too early. (They are called *premature* babies.) They are often very delicate, and may have trouble breathing. To check their breathing, a small plastic bag filled with sponge is taped to the baby's chest. Small movements from breathing compress the air in the sponge. These pressure changes move the thin, flexible diaphragm (metal sheet) in the detector. Because the diaphragm is part of a capacitor the movements can be detected electrically and recorded, so the baby's breathing pattern can be studied.

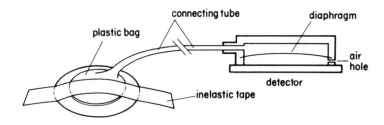

(a) When the baby breathes in the sponge is squashed.
    (i) What happens to the pressure in the bag?
    (ii) Which way does the diaphragm move? *(2)*
(b) Why must the tape fixing the bag to the baby be 'inelastic'? (i.e. be unable to stretch.) *(2)*
(c) Why is there a small air hole in the detector, below the diaphragm? *(2)*
(d) Describe what happens to the metal diaphragm when the baby breathes out again, and the sponge in the bag is not compressed. *(2)*
(e) If the device stopped giving a signal it might be because the baby had stopped breathing. Explain one other way in which the device may stop working, but the baby might be breathing normally. *(2)*

Total 10 marks

73 Prevention of 'cot deaths'

There have been many unexplained deaths of babies. These have been called 'cot deaths'. Sometimes an apparently healthy baby has simply stopped breathing. A special mattress has been made to sound an alarm as soon as the baby stops breathing. The baby lies on the air-filled mattress which is made of separate sections, each with a tube connected to a single detector. As the baby makes slight movements air is forced out of one section into another. To do this it has to flow over the detector. The detector is a thermistor which is heated by a small current flowing through it. As the air blows over it, it changes temperature. The changes of temperature are detected electrically.

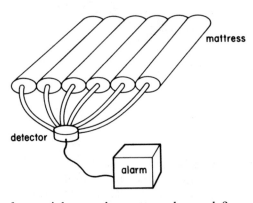

(a) Of what sort of material must the mattress be made? *(1)*

(b)   (i) What can you say about the pressure in each section of the mattress *before* the baby is put on it?

  (ii) What would happen to the pressure in each section if a *stationary* weight was put on it?

  (iii) Why is it that only *movements* cause changes in the detector? *(4)*

(c)   (i) What happens to the temperature of the thermistor when air blows over it?

  (ii) What effect does this temperature change have on the electrical properties of the thermistor? *(3)*

(d) What would eventually happen if a small air leak developed in *one* of the connecting tubes? *(2)*

Total 10 marks

74 Car brakes

A schematic diagram of a car braking system is shown below. The driver presses down on the brake pedal, and this pushes the piston along in the master cylinder. This increases the pressure of the brake fluid in the system. The increased pressures pushes the pistons in the brake cylinders along. The brake cylinders are mounted on each wheel, and when they move they push brake pads hard against a rotating part of the wheel. The friction produced is what slows the car down.

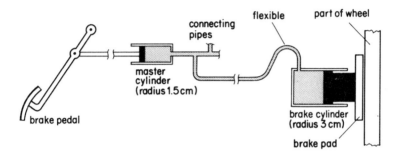

(a) Calculate the area of the piston in the master cylinder. *(1)*
(b) If the brake pedal is pushed so that the force on the piston in the master cylinder is 300 N, calculate the pressure of the fluid. (Ignore any friction.) *(1)*
(c) Why is there a flexible section of pipe connected to the brake cylinder? *(1)*
(d) Calculate the force with which the piston in the brake cylinder pushes on the brake pad. (Again, ignore any friction.) *(2)*
(e) For each millimetre that the piston in the master cylinder moves, how far does the brake pad move? *(1)*
(f) By comparing the arrangement of molecules in a liquid (like brake fluid) and a gas (like air), explain why liquids are difficult to compress, but gases are easy. *(2)*
(g) Why do air bubbles in the brake fluid make the brakes *much* less effective? *(2)*

Total 10 marks

75 Sound in pipes

A power station may contain miles of pipes with water in them. If a leak occurs, or there is some obstruction in a pipe, the repair crew will need to find where the fault is. To do this they empty the pipe and then use a loudspeaker to send a pulse of sound along the pipe, and a small microphone to pick up any echo. Any time delay between the pulse and its echo tells the crew how far away the fault is. The time delay is measured by showing the echo on the screen of a cathode ray oscilloscope (CRO). A simplified diagram of the apparatus and traces on the CRO is shown below.

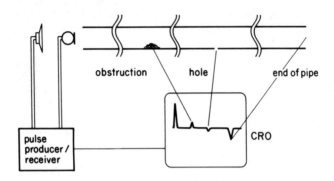

(a) How does the time delay depend on how far away the fault is? *(2)*
(b)   (i) If the speed of sound in a pipe is 350 m/s and the time delay is 2 s, how far away is the fault?
    (ii) Calculate the time delay for a fault which is 175 m away. *(2)*
(c) How could you tell if the pipe was completely blocked? *(1)*
(d) How would the trace be different for different size holes: (Hint: look at the trace for the end of the tube.) *(1)*
(e) In which direction do the air molecules in the pipe move as the sound wave travels along? *(1)*
(f) The pulse produced by the loudspeaker is a sudden *compression* of the air, and this travels along the tube. By looking at the shape of the trace, say what bounces back from
    (i) a blockage
    (ii) a hole
    (iii) the end of the pipe. *(3)*

Total 10 marks

SONAR stands for *SO*und *N*avigation *A*nd *R*anging. Sound waves are sent out from a ship through the water. Objects in the water, like shoals of fish or submarines, reflect some of the sound, and this is picked up by an under-water microphone. Telling how far away the objects are is not simple. The pressure of the water increases as the depth increases, and this speeds up the sound waves. The temperature of sea water is almost constant to a depth of about 150 m, but then it rapidly gets colder. This slows the sound waves down. These changes in speed cause the sound waves to change direction. Some sound waves are reflected from the underneath surface of the sea, or totally internally reflected from the water at greater pressure. Graphs showing the variation of temperature and speed of sound with depth are shown below. The diagram shows possible paths of sound waves.

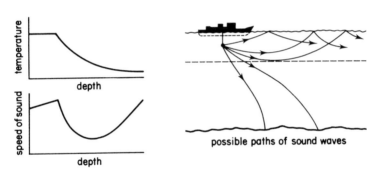

possible paths of sound waves

(a) A pulse of sound is sent out from a ship and an echo returns 10 s later. If the speed of sound through the water is about 1500 m/s, estimate how far away the reflecting object is. *(3)*

(b) Use the passage to say why the word *estimate* is used in (a). *(2)*

(c) Draw diagrams to show sound waves bending as they cross from one layer of water into another where they
   (i) speed up, and (ii) slow down. *(4)*

(d) Use your answers to (c) and the graph of speed against depth to explain why the beam of sound marked X bends first upwards, then downwards. *(4)*

(e) Suggest why there is a layer of water at the surface of the sea which has an almost steady temperature. *(2)*

Total 15 marks

77 Reporting of radioactivity

There have been serious accidents in nuclear energy in recent years, at Three Mile Island in America, Chernobyl in Russia, and various leaks at Sellafield in England. People are understandably worried about the safety of nuclear energy. However, news reports are not always very helpful. They are sometimes misleading because the reporters themselves may not understand the topic very well. The word 'radiation' is used without saying that there are three types, and phrases like 'a cloud of radiation' are used quite wrongly. People may think that *any* amount of radiation from radioactive substances is dangerous, and not know that there has always been some radiation in our atmosphere. Many people may also not know that radiation of this sort is used a great deal in hospitals, laboratories and in industry.

(a) Name the three types of radiation that come from radioactive substances. *(3)*

(b) Write down a correct phrase to replace 'cloud of radiation'. *(2)*

(c) Which sort of radiation is easiest to shield against if the radioactive substance can be put in a container? Explain your answer. *(2)*

(d) Which sort of radiation would be least dangerous to you if you breathed in radioactive dust? Explain your answer. *(2)*

(e) A radioactive source is suspected of giving out beta and gamma radiation. How is it possible to test if this is correct? (There are at least two ways; give one.) *(2)*

(f) What does the term 'half-life' mean? *(1)*

(g) A radioactive substance has a half-life of 2 hours. To what fraction will the radiation be reduced after 6 hours? *(1)*

(h) In an accident in a nuclear power station, there is a leak of a radioactive substance with a half-life of 25 years. The radiation level rises to 8 times the safety level. How long will it be before the power station is safe to enter again, if nothing is done but to wait? *(1)*

(i) What part of an atom does radiation come from? *(1)*

(j) What changes take place in an atom when it gives out each type of radiation? *(3)*

(k) Very briefly, describe *one* way in which radiation is used in *one* of hospitals, laboratories or industry. *(2)*

Total 20 marks

78 Smoke detector using radioactivity

Many fire alarms actually detect smoke. One type uses radioactivity. A weak alpha particle source is used. This may be the element americium 241 (atomic number 95). The source is covered with a thin layer of gold. The number of alpha particles given out is approximately 40 000 each second. The alpha particles cause the air in the smoke detector to be ionised. Each alpha particle can produce about 140 000 pairs of ions. This means that a small current flows through the air from one electrode to the other. If smoke gets into the detector through the 'smoke louvres' many of the ions cling to the tiny particles of ash and dust which make up the smoke. Each ion has therefore effectively increased in mass, and so cannot be attracted to the electrodes so quickly. The current falls, this fall is detected electronically, and sounds the alarm.

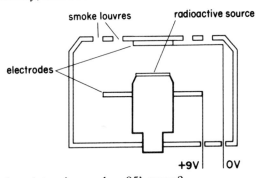

(a) (i) What does 'atomic number 95' mean?
    (ii) What does the number 241 tell you? (2)
(b) What is ionisation? (1)
(c) How many ions are produced each second? (1)
(d) Why is there a layer of gold over the source, and why must it be a *very* thin layer? (2)
(e) Which way do negative ions move through the air between the electrodes? (1)
(f) Which basic law of physics tells you that the ions will not be accelerated so quickly if their mass increases? (1)
(g) Imagine that you are trying to sell this sort of smoke detector, but your customer is worried about the dangers of radioactivity. Give two reasons why the detector is safe. (2)

Total 10 marks

79 Energy from the nucleus

At present, nuclear energy for weapons and nuclear reactors is produced by 'fission'. This is a process of heavy atoms breaking up into smaller fragments. The break up can happen naturally, or can be caused by a neutron colliding with the nucleus of a heavy atom. A typical reaction is shown below:

$$^{235}_{92}U + ^{1}_{0}n \rightarrow ^{244}_{56}Ba + ^{90}_{36}Kr + 2^{1}_{0}n + energy$$

The main trouble with this process is that the 'fission products' like barium (Ba) and krypton (Kr) are radioactive. They may have very long half-lives, making safe disposal a problem.

A different nuclear process called 'fusion' is the combining of light nuclei. A typical reaction is shown below:

$$^{2}_{1}H + ^{2}_{1}H \rightarrow ^{3}_{3}He + ^{1}_{0}n + energy$$

This process has the advantage that the products are not radioactive. However it is only at the experimental stage for peaceful use, because the isotope of hydrogen, called heavy hydrogen or deuterium, must be at a temperature of about 350 000 000 °C before the reaction will happen continuously! (This is the sort of reaction which is going on in the sun.)

(a) Name the three particles which make up most atoms, and say what sort of electrical charge they have. *(3)*

(b) Which two particles make up the nucleus of an atom? *(1)*

(c) What does the atomic number of an element tell you? *(1)*

(d) What sort of electrical charge does the nucleus of an atom have? *(1)*

(e) Why do the nuclei of barium and kripton formed in the first reaction repell each other strongly? *(1)*

(f) Explain why it is easy for the neutron in the first reaction to collide with the uranium nucleus, but extremely difficult for two nuclei of deuterium to collide with each other in the second reaction. *(2)*

(g) How does the extremely high temperature make the second reaction possible? *(1)*

Total 10 marks

80 Gamma radiation

X-rays are very widely used in medicine. A beam of X-rays is shone through a part of the body, and the shadows of bones are recorded on a photograph. The same sort of thing is done in industry, to see if there are hidden faults inside metal castings and joins.

However, X-ray machines are large, costly, and generate a lot of heat. For these reasons, gamma rays are often used instead. They behave in exactly the same way as X-rays, but come from a very different source. A gamma ray source is a small piece of radioactive material, so it is easily portable, it does not get hot, and it is relatively cheap.

Unfortunately there are some snags. Both X-rays and gamma rays are dangerous. An X-ray machine can be switched on only when needed, but a gamma ray source is on all the time. An X-ray machine produces most of the radiation travelling in one direction, but a gamma ray source sends radiation in all directions.

(a) Give two advantages and two disadvantages of gamma rays compared with X-rays, for industrial use. *(4)*

(b) A long pipeline to carry oil across miles of desert is being built. The sections of pipe are welded together. Should an engineer use X-rays or gamma rays to examine the welded joints in the pipe? Explain your choice. *(3)*

(c) In (b) the source is placed inside the pipe, and the photographic film wrapped round the outside. Explain whether it is an advantage or a disadvantage for gamma radiation to be given out in all directions. *(2)*

(d) The film mentioned above is quite ordinary photographic film.
   (i) Explain why the film is wrapped in thick black paper.
   (ii) Explain why it does not have to be unwrapped when gamma radiation is used to take a 'photograph'. *(2)*

(e) As gamma radiation is on all the time, it is controlled by blocking it off when it is not needed.
   (i) Name a suitable material to block off the radiation.
   (ii) Sketch a design for a holder which will block off the radiation when it is not needed, but can be opened to let out a beam of radiation in one direction only. *(4)*

Total 15 marks

81 Disposal of radioactive waste

A nuclear reactor produces cheap energy, but also produces radioactive waste. Radioactive substances can give out three types of radiation, alpha particles, beta particles and gamma radiation. All three of these are very dangerous to anything living. Alpha particles can only travel short distances in air, and can be stopped with just a sheet of paper. Beta particles can travel further, about one metre in air, but can still be stopped easily, this time by a few millimetres of metal. Gamma radiation is much more penetrating; it is hardly affected at all by air. It is cut down, but not stopped entirely, by very thick sheets of lead. Lead is expensive, so even thicker barriers of steel and concrete are used.

The danger from radioactivity does decrease, because the amount of radiation decreases. Unfortunately the half-life of many radioactive substances is very long. It is therefore many years, maybe even thousands, before some waste may be safe.

Various proposals have been made for safe disposal of the waste. The waste could be sealed in steel drums or glass or concrete blocks. It could then be kept at power stations or dumped at sea. It could be buried in the ground, either close to the surface or deep down in stable rocks.

(a)   (i) Which type of radiation is the most difficult to stop?
     (ii) Which is the easiest to stop?   (2)

(b) If some waste only gave out alpha and beta particles all the radiation could be stopped by putting it in a steel drum a few millimetres thick. Why would this *not* be a safe thing to do, if it then has to be stored for many years or dumped?   (3)

(c) Which method of disposal mentioned in the passage would be furthest away from any living creatures?   (2)

(d) Which method of disposal mentioned in the passage do you think would be most expensive?   (2)

(e)   (i) What does 'half-life' mean?
     (ii) Some waste has a half-life of 30 years. What fraction will be left after 60 years?   (4)

(f) Do you think that nuclear energy can be made safe enough to use? Give one reason for your answer.   (2)

Total 15 marks

82 Thickness gauge

When thin sheet materials are being made, the maker needs to be sure that the thickness is correct. The diagram below shows a beta particle thickness gauge used in measuring thin aluminium sheet. Some of the beta particles are absorbed as they go through the sheet. The number of beta particles counted by the detector gives a measure of the thickness. The number varies randomly even when the thickness is constant, so the detector has to take a count averaged over several seconds.

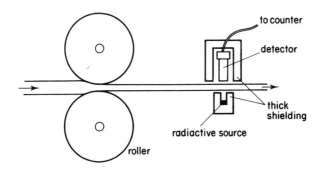

(a) If the aluminium gets thicker, what happens to the number of particles counted? *(2)*

(b) What should the machine operator (or the automatic controls) do to correct the situation in (a)? *(2)*

(c) What does 'The number (of beta particles) varies randomly . . .' mean? *(2)*

(d) A lot of dust in the atmosphere might absorb some beta particles. In what way would this make the gauge give a wrong reading? *(2)*

(e) Beta particles are dangerous. How would you explain to a worker in this factory that the thickness gauge has been made safe? *(3)*

(f) There are three types of radiation from radioactive substances, alpha, beta and gamma radiation. Explain why neither alpha nor gamma radiation could be used for this thickness gauge. *(4)*

Total 15 marks

Answers to numerical questions

1 (b) 0.6 W (c) 0.33 A (d) 15Ω,36Ω (e)(i) £4 (f)(i) 0.4 A,2.7 A
(ii) × 6.7

2 (d)(i) ×2 (ii) ×2 (iii) ×2 (iv) × ¼ (v) ×2.8

3 (a)(i) 1.47 p (ii) 0.36 p (e)(i) 5.3 p (ii) 32 p (iii) 0.66 p (f) 21 p

4 (b)(i) × ½ (ii)4

5 (d)(ii) 0.33 A (iii) 1.7 A

6 (b)(i) 60% (ii) 70% (i)(i) 400 000 J (ii) 300 000 J

9 (a)(i) 6 V (ii) 4.5 V (iii) 0 V

20 (d) 5 J (e) 720 000 kg

23 (a) 12 000 V

25 (a) 0 V (b) 10 V

30 (a) 2 V (c) 0.8 s (d) 1.25 Hz (e) 1 A

31 (a) 60 (c) 30 (e) 500 Hz

35 (a) 2 °C/min (e)(i) 210 000 J (ii) 7000 J

37 (c) 7 cm (f) −280 °C

38 (e)(i) 38.4 W (ii) 38.4 J (iii) 4608 J (iv) 23 040 J (v) 4608 J
(h) 1400 J/ kg °C

41 (a) 30% (d)(i) 1500 °C (ii) 69 000 000 J (iii) 27 500 000 J
(iv) 96 500 000 J (v) 138 000 000 J

50 (a) 30 cm (e) 6.5 h approx.

56 (d) 6300 nm

57 (a)(i) 100 min (ii) 60 km/h (b)(i) 50 km (ii) 1 h 45 min (iii) 29 km/h
(c) 50 km/h, 90 km/h, 40 km/h (d) 100 km/h (e)(i) 1000 km (ii) 20 h
(iii) 50 km/h (f)(i) 20 m/s (ii) 40 s (iii) 0.5 m/s per s (iv) 8000 N

58 (c)(i) 3 000 000 kg m/s (ii) 4500 MJ (e)(i) 2 m/s^2 (ii) 200 m/s

59 (a) 800 N (b) 640 J (d) 900 J

60 (a) 16 m (b) 10 s (c) 0.8 m/s (e) 600 J (f) 9600 J (g) 9600 W (i)(i) 96 J
(ii) 1536 J

61 (a) 5000 N (b)(i) 1000 kg m/s (ii) 1000 J (c) 1000 J (d) 20 cm (e) 2 m/s
(f)(i) 1 m/s (ii) 0.05 s (g) 20 000 N

62 (b)(i) 5 m (ii) 3500 J (iii) 10 m/s (iv) 1 s (c)(i) 5 m/s (ii) 0.2s
(iii) 700 kg m/s (iv) 3500 N

63 (a) 8 N (b) 0.25 cm (c) 80 m/s^2,10m/s^2 (d) 12 N (e) 120 m/s^2 (g)(i) 3 V
(ii) 6 V (iii) 0 V

64 (a)(i) 400 J (ii) 400 J (iii) 6.3 m/s (d)(i) 126 kg m/s (ii) 12 600 N
(iii) 1260 N

65 (a) 0.1 m/s^2 (e)(i) 0 N (ii) 200 000 N (f)(i) 0.15 m (ii) 0.75 m

66 (f)(i) 1064 N (ii) 364 N

67 (a) 15 m/s (b) 10 s (c) 2 s (d)(i) 11 000 kW (e) 88 000 m

68 (d)(ii) 50 J

69 (a)(i) 9000 kg m/s (ii) 0 kg m/s (iii) 9000 kg m/s (iv) 18 000 N
(v) 2000 N
74 (a) 7.07 cm^2 (b) 42.4 N/cm^2 (d) 1200 N (e) 0.25 mm (assuming one brake cylinder only!)
75 (b)(i) 350 m (ii) 1 s
76 (a) 7500 m
77 (g) ⅛ (h) 75 yr
78 (c) 5600 million
81 (e)(ii) ¼

ISBN 0 340 42175 4

First published 1987
Second impression 1988

Copyright © 1987 A. York

Typeset by Photoprint, Torquay, Devon
Printed in Great Britain by Richard Clay Ltd, Bungay, Suffolk
for Hodder and Stoughton Educational
a division of Hodder and Stoughton Ltd, Mill Road,
Dunton Green, Sevenoaks, Kent